ANALOG VLSI IMPLEMENTATION OF NEURAL SYSTEMS

THE KLUWER INTERNATIONAL SERIES IN ENGINEERING AND COMPUTER SCIENCE

VLSI, COMPUTER ARCHITECTURE AND DIGITAL SIGNAL PROCESSING

Consulting Editor
Jonathan Allen

Other books in the series:

Logic Minimization Algorithms for VLSI Synthesis. R.K. Brayton, G.D. Hachtel, C.T. McMullen, and A.L. Sangiovanni-Vincentelli. ISBN 0-89838-164-9.
Adaptive Filters: Structures, Algorithms, and Applications. M.L. Honig and D.G. Messerschmitt. ISBN 0-89838-163-0.
Introduction to VLSI Silicon Devices: Physics, Technology and Characterization. B. El-Kareh and R.J. Bombard. ISBN 0-89838-210-6.
Latchup in CMOS Technology: The Problem and Its Cure. R.R. Troutman. ISBN 0-89838-215-7.
Digital CMOS Circuit Design. M. Annaratone. ISBN 0-89838-224-6.
The Bounding Approach to VLSI Circuit Simulation. C.A. Zukowski. ISBN 0-89838-176-2.
Multi-Level Simulation for VLSI Design. D.D. Hill and D.R. Coelho. ISBN 0-89838-184-3.
Relaxation Techniques for the Simulation of VLSI Circuits. J. White and A. Sangiovanni-Vincentelli. ISBN 0-89838-186-X.
VLSI CAD Tools and Applications. W. Fichtner and M. Morf, editors. ISBN 0-89838-193-2.
A VLSI Architecture for Concurrent Data Structures. W.J. Dally. ISBN 0-89838-235-1.
Yield Simulation for Integrated Circuits. D.M.H. Walker. ISBN 0-89838-244-0.
VLSI Specification, Verification and Synthesis. G. Birtwistle and P.A. Subrahmanyam. ISBN 0-89838-246-7.
Fundamentals of Computer-Aided Circuit Simulation. W.J. McCalla. ISBN 0-89838-248-3.
Serial Data Computation. S.G. Smith and P.B. Denyer. ISBN 0-89838-253-X.
Phonologic Parsing in Speech Recognition. K.W. Church. ISBN 0-89838-250-5.
Simulated Annealing for VLSI Design. D.F. Wong, H.W. Leong, and C.L. Liu. ISBN 0-89838-256-4.
Polycrystalline Silicon for Integrated Circuit Applications. T. Kamins. ISBN 0-89838-259-9.
FET Modeling for Circuit Simulation. D. Divekar. ISBN 0-89838-264-5.
VLSI Placement and Global Routing Using Simulated Annealing. C. Sechen. ISBN 0-89838-281-5.
Adaptive Filters and Equalisers. B. Mulgrew, C.F.N. Cowan. ISBN 0-89838-285-8.
Computer-Aided Design and VLSI Device Development, Second Edition. K.M. Cham, S-Y. Oh, J.L. Moll, K. Lee, P. Vande Voorde, D. Chin. ISBN: 0-89838-277-7.
Automatic Speech Recognition. K-F. Lee. ISBN 0-89838-296-3.
Speech Time-Frequency Representations. M.D. Riley. ISBN 0-89838-298-X
A Systolic Array Optimizing Compiler. M.S. Lam. ISBN: 0-89838-300-5.
Algorithms and Techniques for VLSI Layout Synthesis. D. Hill, D. Shugard, J. Fishburn, K. Keutzer. ISBN: 0-89838-301-3.
Switch-Level Timing Simulation of MOS VLSI Circuits. V.B. Rao, D.V. Overhauser, T.N. Trick, I.N. Hajj.
ISBN 0-89838-302-1
VLSI for Artificial Intelligence. J.G. Delgado-Frias, W.R. Moore (Editors). ISBN 0-7923-9000-8.
Wafer Level Integrated Systems: Implementation Issues. S.K. Tewksbury. ISBN 0-7923-9006-7
The Annealing Algorithm. R.H.J.M. Otten & L.P.P.P. vàn Ginneken. ISBN 0-7923-9022-9.
VHDL: Hardware Description and Design. R. Lipsett, C. Schaefer and C. Ussery. ISBN 0-7923-9030-X.
The VHDL Handbook. Dr. Coelho. ISBN 0-7923-9031-8.
Unified Methods for VLSI Simulation and Test Generation. K.T. Cheng and V.D. Agrawal. ISBN 0-7923-9025-3
ASIC System Design with VHDL: A Paradigm. S.S. Leung and M.A. Shanblatt. ISBN 0-7923-9032-6.
BiCMOS Technology and Applications. A.R. Alvarez (Editor). ISBN 0-7923-9033-4.

ANALOG VLSI IMPLEMENTATION OF NEURAL SYSTEMS

edited by

Carver Mead
California Institute of Technology

and

Mohammed Ismail
Ohio State University

KLUWER ACADEMIC PUBLISHERS
Boston/Dordrecht/London

Distributors for North America:
Kluwer Academic Publishers
101 Philip Drive
Assinippi Park
Norwell, Massachusetts 02061 USA

Distributors for all other countries:
Kluwer Academic Publishers Group
Distribution Centre
Post Office Box 322
3300 AH Dordrecht, THE NETHERLANDS

Library of Congress Cataloging-in-Publication Data

Analog VLSI implementation of neural systems.

(The Kluwer international series in engineering and computer science ; SECS 80)
"Proceedings of a workshop . . . held May 8, 1989 in connection with the International Symposium on Circuits and Systems"—Foreword.
Includes index.
1. Neural computers—Congresses. 2. Integrated circuits—Very large scale integration—Congresses. I. Mead, Carver. II. Ismail, Mohammed. III. International Symposium on Circuits and Systems (1989 : Portland, Or.) IV. Series.
QA76.5.A44 1989 006.3 89-15452
ISBN 0-7923-9040-7

Printed in the United States of America

Contents

FOREWORD

This volume contains the proceedings of a workshop on Analog Integrated Neural Systems held May 8, 1989, in connection with the International Symposium on Circuits and Systems. The presentations were chosen to encompass the entire range of topics currently under study in this exciting new discipline. Stringent acceptance requirements were placed on contributions: (1) each description was required to include detailed characterization of a working chip, and (2) each design was not to have been published previously. In several cases, the status of the project was not known until a few weeks before the meeting date. As a result, some of the most recent innovative work in the field was presented. Because this discipline is evolving rapidly, each project is very much a work in progress. Authors were asked to devote considerable attention to the shortcomings of their designs, as well as to the notable successes they achieved. In this way, other workers can now avoid stumbling into the same traps, and evolution can proceed more rapidly (and less painfully).

The chapters in this volume are presented in the same order as the corresponding presentations at the workshop. The first two chapters are concerned with finding solutions to complex optimization problems under a predefined set of constraints. The first chapter reports what is, to the best of our knowledge, the first neural-chip design. In each case, the physics of the underlying electronic medium is used to represent a cost function in a natural way, using only nearest-neighbor connectivity.

Chapters 3 and 4 are concerned with sophisticated nonlinear processing of time-domain signals. In both cases, this processing is carried out in real time, with only a small expenditure of energy per unit computation.

Chapters 5 and 6 describe two of the many projects currently under way to create electronic "neural networks" of the kind often modeled on digital systems. The success of these and other programs focused on the same goal will expand by many orders of magnitude the range of problems accessible to neural network solutions.

Chapters 7 through 10 contain reports of self-contained system chips that perform various kinds of image processing. In each case, the chip contains its own array of phototransducers; the input signals are extracted directly from an optical image focused directly on the chip's surface. Each project is directed at a particular aspect of image analysis. Each is, in its own way, inspired by the organization of the visual system of higher animals.

In aggregate, these chapters give a remarkable portent of things to come. It is clear that the continued evolution of this technology will produce systems possessing characteristics that emulate many of the remarkable properties observed in living systems, but that we have been unable to attain using existing engineering techniques.

Carver Mead
Mohammed Ismail

ANALOG VLSI IMPLEMENTATION OF NEURAL SYSTEMS

1

A Neural Processor for Maze Solving

Christopher R. Carroll

Computer Engineering
University of Minnesota, Duluth
Duluth, Minnesota 55812

This paper describes an nMOS integrated circuit designed in the late 1970's that performed the computationally expensive portion of a maze-solving algorithm using a fine-grained parallel processor architecture. The algorithm included continuously variable weights associated with travel through the maze in different directions. The integrated circuit described here directly incorporated those weights as analog parameters affecting inter-processor communication of digital data. The combination of fine-grained parallelism and inter-processor communication controlled by analog weights was unique, and can be viewed as an early example of what might now be called a neural system.

INTRODUCTION

In the late 1970's, the processing power available from VLSI technology was just beginning to be recognized. Researchers were exploring many different approaches to the technology in an attempt both to use that processing power efficiently and to cope with the complexity that is inherent in circuits at the VLSI level. Many buzz words developed, each referring to a different approach to the problem of designing useful functions within this complexity. Smart memories, systolic arrays, array processors, etc. all had their proponents, and some of these approaches have led to continuing topics for research.

This paper describes an nMOS integrated circuit design that originated as an example of fine-grained parallel processing, and developed into something that today might be recognized as an early example of what is now called neural processing. The chip's purpose was to perform the computationally expensive part of a maze-solving algorithm, using a fine-grained parallel processor architecture. The goal of this paper is to explain how decisions faced during the design led to the unique circuitry that justifies calling this chip a neural system.

In the sections that follow, the development and design of this early neural system will be traced. First, in order to properly motivate the discussion, the basic maze-solving algorithm implemented in the chip is presented. The following section then details the design and implementation of a predecessor chip that solved mazes without the benefit of neural techniques. Next, an extension to the maze-solving algorithm is presented, followed by a discussion of the design of the neural processor chip that dealt with that extension. Finally, some lessons learned from an evaluation of the design and performance of the chip are presented, followed by some conclusions.

THE MAZE-SOLVING ALGORITHM

The maze-solving algorithm selected for implementation in hardware was proposed by E. Moore [5] and extended by C.Y. Lee [4], and again by S. Akers [1]. It is a scheme for finding the shortest route between two points in a plane, where the route is composed of some number of orthogonal line segments through a rectangular array of cells superimposed on the plane. The cell-to-cell spacing, or pitch of the array, equals the width of the path, and movement along the path is restricted to be only between cells that are adjacent to the north south, east, or west. Walls in the maze are created by blocking some of the cells in the array, preventing passage through those cells.

The algorithm finds the shortest path between two cells in the array in two phases. Starting at one endpoint of the path, the first phase, or *propagation phase*, distributes throughout the array of cells information telling how to get back to the original endpoint from each of the other cells. The second phase, or *retrace phase*, then uses that information to find the required path.

The operation of the propagation phase of the algorithm can be visualized by imagining a wavefront of activity expanding out from the original path endpoint much like a ripple in a pond caused by a thrown stone. As the wavefront passes each cell in the array, information is stored in that cell recording from which direction the wavefront approached. This stored information in the cells can be thought of as arrows pointing back to the origin of the propagating wavefront. In the event of the wavefront reaching a cell simultaneously from, say, the south and east directions, both a south-pointing and an east-pointing arrow should be stored in the cell to properly record the options available for finding the way back to the original cell. The propagating wavefront does not penetrate blocked cells, and must distort when such obstacles are encountered. Eventually, if a path exists between the specified endpoints, the propagation will reach the second endpoint of the path, and arrows will be stored indicating the direction to take from that point to find the shortest route to the original endpoint.

The operation of the retrace phase is obvious once the propagation phase has filled the array of cells with information represented by arrows. The retrace phase merely starts at the second path endpoint, reads the information there, and proceeds in the direction indicated by the arrow to a neighbor cell. Once there, it follows the arrow stored in that cell to the next neighbor, and then proceeds one cell at a time following arrows back to the original endpoint. No further modifications to the information stored in the cells are required.

Clearly, the computationally expensive part of this algorithm for maze-solving lies in the propagation phase. That part of the algorithm stores information in a number of cells that is related quadratically to the path length.

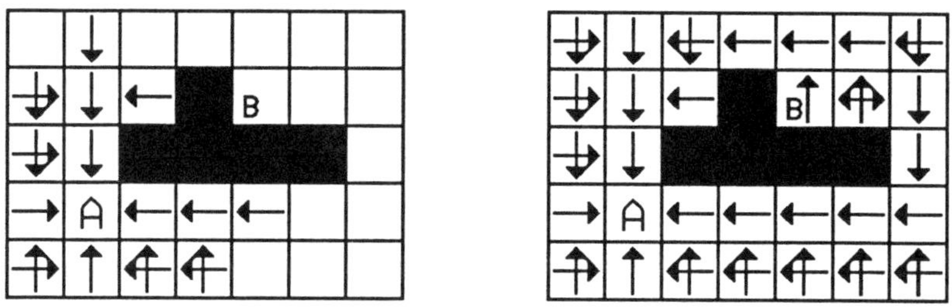

Figure 1 - The propagation phase in progress (left), and finished (right)

Figure 1 shows the operation of this phase of the algorithm. The retrace phase touches only cells along the path, resulting in a linear execution time with path length. Thus, the hardware to be discussed in the following sections attacks only the propagation phase of this algorithm and leaves the computationally easy retrace phase to be executed by a traditional host processor to which this hardware would be attached.

THE *MAZER* CHIP

A good way to relate the propagation phase algorithm described above to the physical world is to imagine that the array of cells is an array of mousetraps, each cocked and ready to fire [6]. Along with each mousetrap is a mechanism that causes it to fire whenever any of its neighbors fires, and a recording device to note from which direction(s) the firing signal comes. With all the mousetraps cocked, propagation is started at one endpoint of the required path by triggering the mousetrap at that location. The mechanisms that link neighboring cells then spread information throughout the array by propagating the wavefront in an expanding frontier of activity to the edges of the array. When complete, a record of the direction taken by the passing wavefront is left in each cell of the array.

Figure 2 is a conceptual logic design of a simplified electronic mousetrap cell. Not shown are all the mechanisms for accessing the information stored in the cell from the host processor, for blocking this cell so that it becomes part of a wall in the maze, or for causing this cell to be the starting point of wavefront propagation, but the mousetrap characteristic is illustrated. After the *reset* line has gone high to make all the latch outputs low, all signals that cross the cell boundary are low, and the system is stable in this state, with all mousetraps cocked. Now, if one of the incoming signals goes high, the corresponding latch will be set. This causes the inputs to the other latches to be disabled via the AND gates, and also causes the cell to generate a high going signal to each of its neighbors, triggering them in the same way. The latches remember from which direction the activation signal entered the cell, and reading them out by an accessing mechanism not shown gives the direction the maze solution takes as it passes through this cell.

An nMOS integrated circuit named *Mazer* was designed in 1977 that implemented a four by four array of cells similar to the one described above, but included the required circuitry to allow the needed interaction from a generic host processor for controlling the start of propagation and for accessing information stored in the latches. The chip included bonding pads conveying to the outside world the propagating signals from cells around the periphery of the four by four array, so that multiple chips could themselves be assembled into an array, expanding the size of the maze that could be solved. The result was essentially a very fine-grained parallel processing system, with each cell's circuitry

---

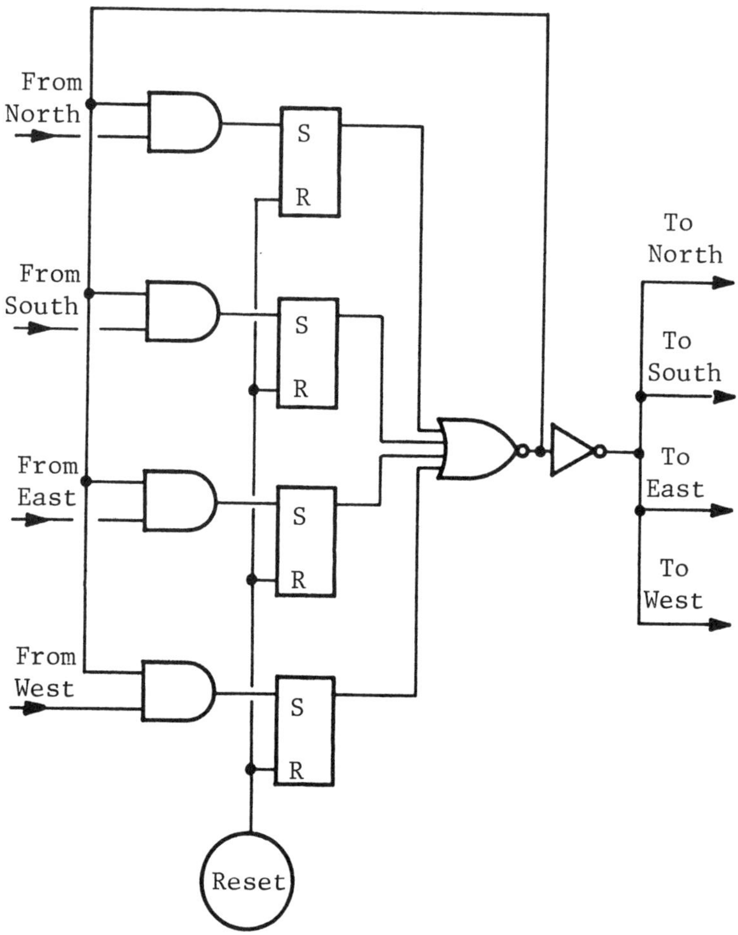

Figure 2 - An electronic mousetrap

representing a processor embedded in a grid of other identical processors. These processing nodes, although consisting of only about twenty gates each, nevertheless performed identifiable tasks of data computation and communication, and thus could truly be called processors. The chip was fabricated and tested in 1978.

In the course of testing the *Mazer* chip, an interesting anomaly showed up. In situations such as that depicted in Figure 3, where a path was to join cells located in different *Mazer* chips in a multi-chip array, unexpected results occurred, such as that shown in the figure. A thoughtful analysis of the situation revealed that the *Mazer* system was not finding paths based on the shortest distance between cells, but rather based on the shortest propagation time of the propagating wavefront between cells. Because the wavefront propagated much more quickly between adjacent processors that were on the same *Mazer* chip than it did between adjacent processors that happened to be on different chips, the chip boundaries in the array of cells established artificial barriers which, though crossable, imposed a high penalty on a path that traversed them. Thus a path between processors on different chips was chosen by this system more on the basis of how many chip boundaries needed to be crossed than on the total path length, resulting in the type of anomalies displayed in Figure 3. This effect played an important role in the design of the chip to be described next.

TWO-LAYER PATH FINDING

The fact that the *Mazer* was restricted to solving mazes embedded in a plane limited its usefulness. The most immediate application for path-finding hardware was in the area of wire routing on printed circuit boards or on silicon chips, but for such applications at least two levels of wiring were required for reasonable wiring efficiency. Thus, there was great incentive to develop a two-layer path finder, with the specific goal of producing a machine capable of

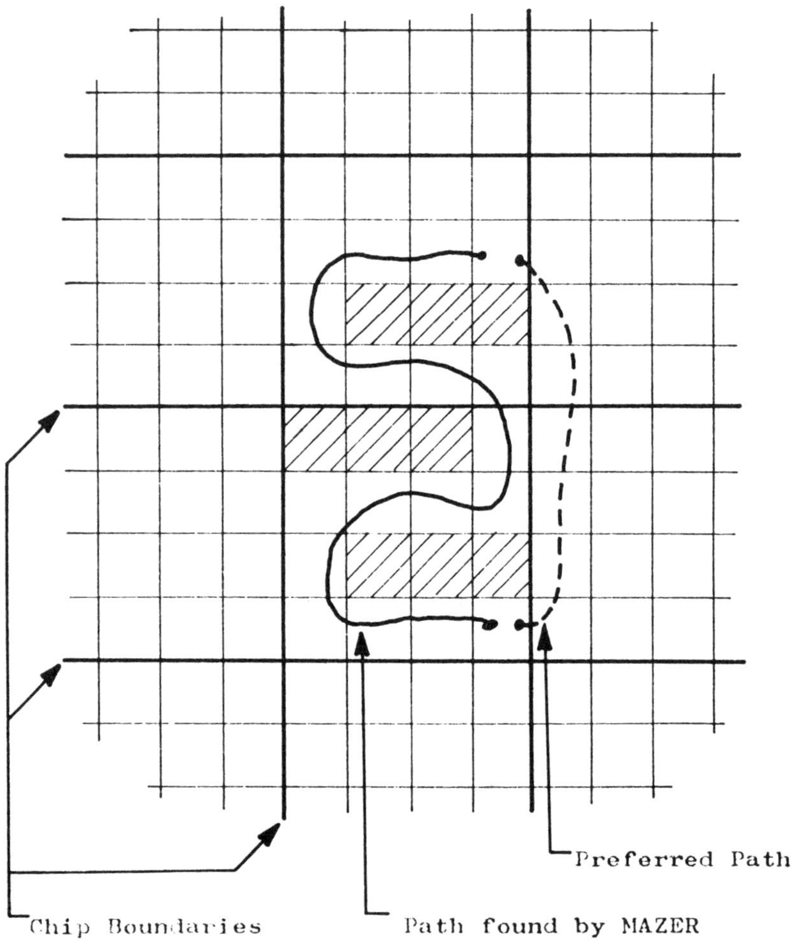

Figure 3 - An anomaly in *Mazer's* operation

routing wires in the applications mentioned above. This was the background that prompted the design of the second integrated circuit to be discussed, the *Pathfinder*.

At first glance, it seemed that one could simply construct a circuit that formed the topology of two *Mazer* chips laid on top of one another, with an additional arrow bit in each cell to indicate travel from one layer to the other. This strategy would have worked, except that it lacked an important property that was needed in the envisioned applications of this system, printed circuit board wire routing.

Designers of two-layer circuit boards have long realized that it is best for mostly vertical wire runs to end up on one side of the board, and mostly horizontal runs to end up on the other side. This helps to avoid unnecessarily blocking channels for future wires. The tendency of a wire to choose one side of the board or the other depending on its orientation would have been completely lacking in a straightforward two-layer *Mazer*. Incorporating this preference into the basic path-finding algorithm was an interesting problem.

A way to achieve the wire location preference was to use a system of costs associated with travel from cell to cell through the array. A mechanism was needed to make travel in some directions more expensive than travel in other directions. With such a mechanism in place, north-south travel could be encouraged on one layer of the maze and east-west travel encouraged on the other layer by making travel in the orthogonal directions on each layer more expensive. A separate cost could be added for travel from one layer to the other, since such travel was often limited in the applications envisioned for this system.

Interestingly, an accidental example of an imposed path cost had already been seen in the *Mazer* system. Crossing chip boundaries with a path connecting cells in different chips of a multi-chip array imposed additional costs on such paths, making routes that crossed fewest chip boundaries preferable over other routes, as discussed in the previous section. This effect resulted from the additional delay imposed on the propagation of the information wavefront between adjacent cells separated by chip boundaries over that between adjacent cells on the same chip. By appropriately controlling the speed with which the propagating wavefront of activity traveled through individual cells in different directions, any desired set of costs could be imposed on the resulting paths. Based on this idea of controlling the speed of wavefront propagation from cell to cell, the second chip design, the *Pathfinder*, was proposed.

THE *PATHFINDER* CHIP

The proposed *Pathfinder* chip would implement a three-cost system for choosing desired paths between specified endpoints in a two-layer maze. The lowest cost would be imposed on east-west travel on the top layer of the maze and north-south travel on the bottom layer. A second, higher cost would be imposed on north-south travel in the top layer and east-west travel on the bottom. A third, still higher cost, would be charged for inter-layer travel. Each of these costs would be implemented with a variable weight that controlled the speed of propagation of the expanding wavefront of activity in the direction assigned to that weight.

Figure 4 shows the effect of controlling wavefront speeds in propagating arrow information through just the top layer of a two-layer maze. Here propagation in the east-west direction was allowed to proceed at a rate three times that of north-south propagation, encouraging east-west paths on this layer of the maze. Only the arrows associated with travel on this top layer are shown, for clarity. As the figure shows, the path indicated by the arrows stored between cells was not the physically shortest path available, but was the least costly path based on the three-to-one ratio of imposed weights. In the two-layer environment, if travel was less costly in the east-west direction on the top layer

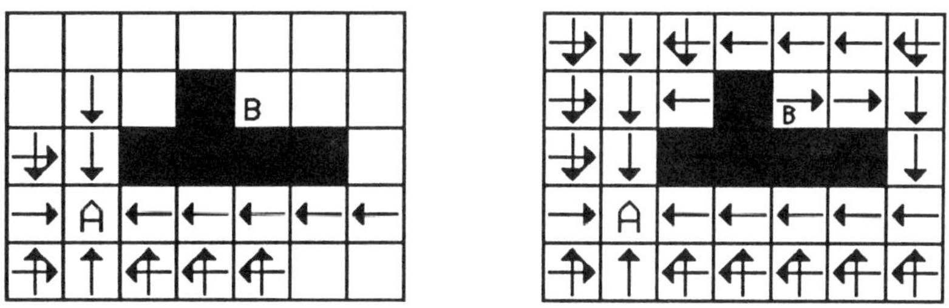

Figure 4 - Propagating three times faster east-west than north-south, in progress (left), and finished (right)

and on the north-south direction on the bottom layer, then propagating wavefronts from a cell A to a cell B that was mostly east of cell A tended to reach cell B more quickly on the top layer, resulting in arrows stored in cell B that indicated a path back to cell A using a route on the top layer. Similarly, cells that were mostly north or south of each other tended to be connected by paths routed on the bottom layer of the maze. This scheme accomplished the desired separation of north-south and east-west paths on different layers of the maze. Paths that used both layers of the maze were possible but less likely because of the higher cost imposed on inter-layer travel.

Some additional design changes were included in the *Pathfinder* that distinguished it from the *Mazer* design. An additional bit of storage was included in each cell to allow blocking travel between layers independently from blocking intra-layer travel through that cell, because the applications envisioned for this system often required such a capability. Also the visualized position of the

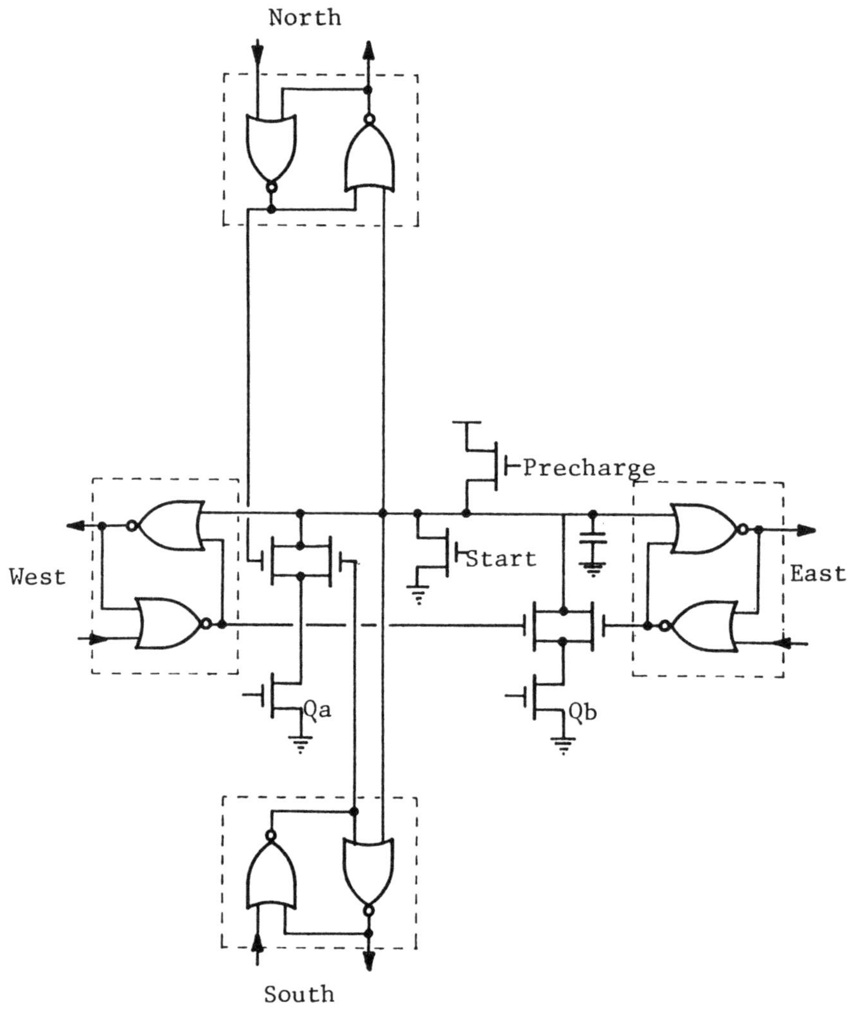

Figure 5 - Simplified one-layer *Pathfinder* processor

arrows stored with each cell was moved from within the cell to between cells. The arrows thus recorded the direction from which the propagating wavefront came as it crossed cell boundaries. This reduced the required storage for the arrow latches by a factor of two, and aided in the interpretation of the information stored in them.

The method employed in the *Pathfinder* for achieving the controllable propagation speeds relied heavily on the dynamic charge storage abilities of MOS circuitry. Figure 5 shows a circuit representing a simplified, one-layer cell with its surrounding arrow latches, but does not show the blocking or accessing circuitry. Each cell contained a capacitor of about 5 pF. Before the start of the propagation phase, the capacitors were all precharged by means of the precharge transistor. With all the capacitors charged, all the arrow latches had both outputs held low. To start propagation at a particular cell, that cell's capacitor was discharged. That action released one side of the arrow latches surrounding that cell, causing those arrows to "point" to that cell with the discharged capacitor. The high outputs of the arrow latches then entered the neighbor cells, and began discharging the capacitors there at rates determined by the voltages on the gates of transistors Qa and Qb. When those capacitors were completely drained, the arrows surrounding those cells flipped to point to the newly discharged capacitors, and the arrow latch outputs began discharging capacitors in their neighbors. As the wavefront of activity propagated out, cells behind the frontier had completely discharged capacitors, cells ahead of the frontier had fully charged capacitors, and cells on the frontier had capacitors that were in the process of being discharged.

The time required, and thus the cost, for propagating through a cell depended on the rate at which the capacitor was discharged, which, in turn, depended on the voltages on the gates of transistors Qa and Qb. The direction from which the wavefront approached the cell determined whether the current path controlled by Qa or the current path controlled by Qb was used in discharging the capacitor.

A feature included on the *Pathfinder* chip allowed a small amount of local control over the cost function to modulate the overall three costs described above. This consisted of an additional 1 pF of capacitance that could be switched on in parallel with the main capacitor in each cell. The time for propagating through a cell, and hence its propagation costs, could be increased by connecting its extra capacitor before precharge and leaving it connected through propagation. The cost could be decreased by connecting the extra capacitor after precharge was over and disconnecting it again before propagation started. These capacitor connections were switched on a cell-by-cell basis, controlled by an additional bit in each cell. This made it possible to increase costs locally in the maze so that paths would tend to avoid certain congested or otherwise undesirable parts of the maze, or to decrease costs to encourage utilization of remote parts of the maze.

Figure 6 is a schematic of a two-layer *Pathfinder* processor, containing circuitry for both layers of the maze and the arrow between them. The north and east arrows for each layer of the cell are arbitrarily shown as a part of this processor, while the arrows to the south and west are considered to belong to the neighbor processors in those directions. The control storage bits are shown as boxes for clarity. Actually the five arrow bits and the four control bits make up a nine-bit word of what amounts to a standard static memory system, using the usual six-transistor cell. Not shown in the figure are the mechanisms that allow the host processor to read and modify these information bits.

The circuit shown in Figure 6 worked just as described above for Figure 5, with the addition of the blocking controls and the addition of the second-layer circuitry. Having two layers simply meant that three current paths were present

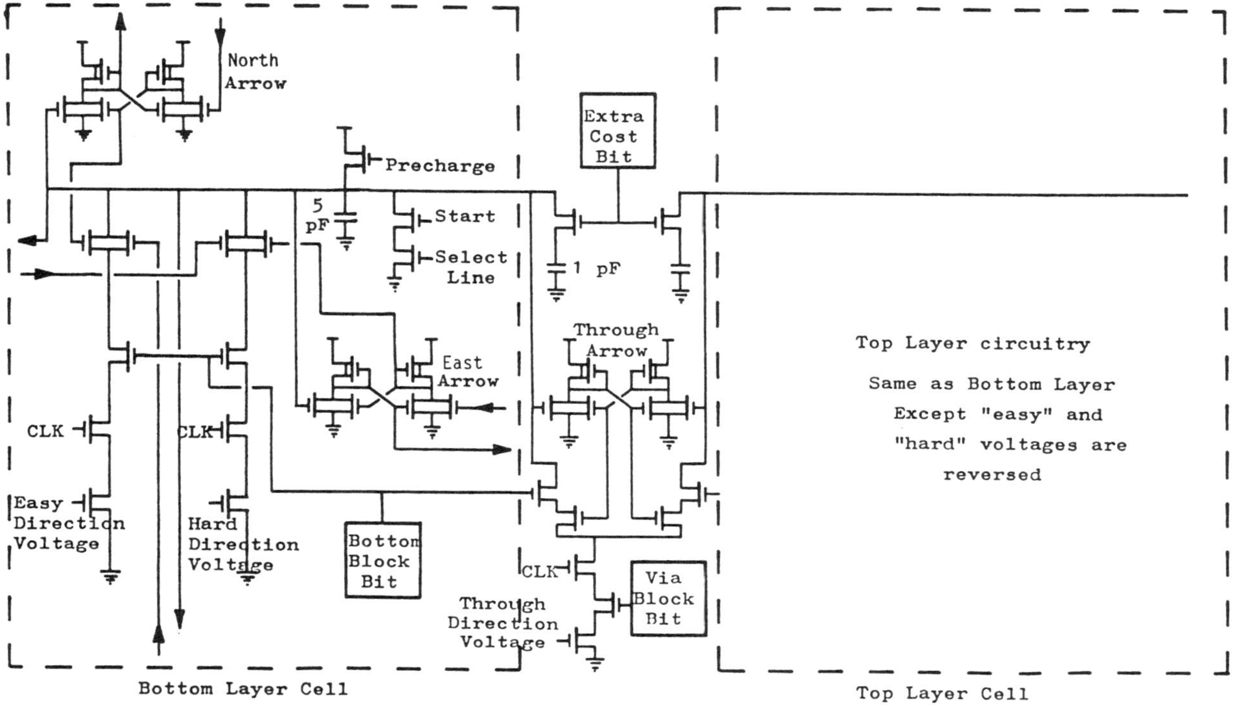

Figure 6 - Schematic of Pathfinder processor

for discharging the capacitors, each controlled by a transistor whose gate voltage determined how quickly the capacitor was discharged through that path. The blocking control latches simply opened the appropriate discharge paths to prevent the discharge of the capacitor under the conditions that were to be blocked. The additional signal labeled CLK in Figure 6 was present to allow the discharging action to be interrupted in the entire array of processors to allow starting propagation at multiple cells simultaneously, or to perform other experiments with the circuit.

Figure 7 shows a plot of the metal layer of the *Pathfinder* chip. The chip contained a four by eight array of two-layer cells. As with the *Mazer*, the large cell arrays needed for useful maze-solving applications were built up by assembling *Pathfinder* chips themselves in an array. Forty-eight of the seventy bonding pads were devoted to chip-to-chip communication within the multi-chip array. The *Pathfinder* was fabricated by MOSIS in 1980.

THE *PATHFINDER* AS A NEURAL SYSTEM

Although the *Pathfinder* chip was designed ten years ago, viewed from today's perspective it displays many of the characteristics associated with what are now called neural systems. Some investigation of those characteristics as displayed in such an early example of this field might reveal insights that could be useful in new designs.

First and foremost, the *Pathfinder* is an example of a parallel processing system with very fine-grained parallelism. This is one characteristic of neural systems, which rely on large numbers of very simple processors to perform computations in the same way that biological systems rely on large numbers of

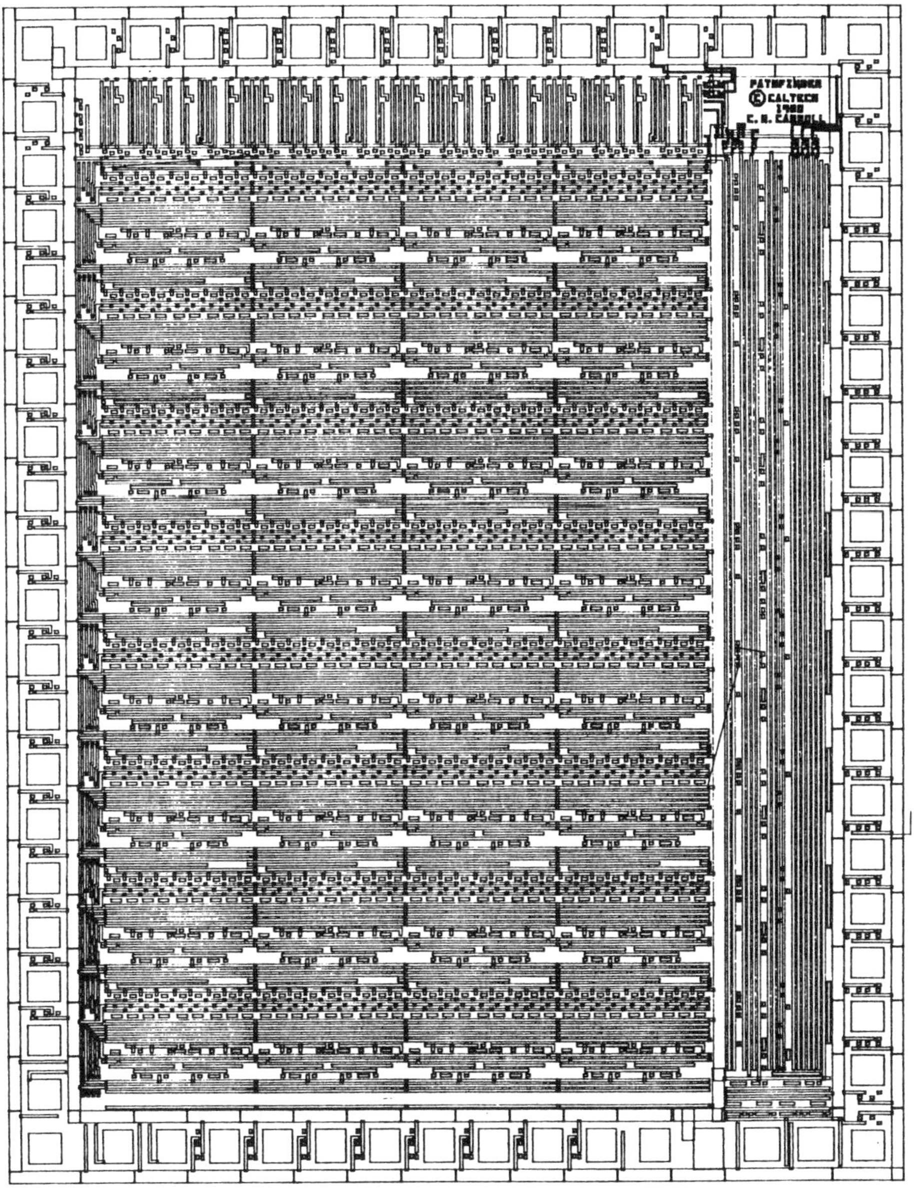

Figure 7 - Plot of Pathfinder's metal layer

simple neurons. In the *Pathfinder*, the processing nodes consist of a capacitor and just a few logic gates, but by networking many of those nodes into a system, useful computation results.

A second characteristic of neural systems is a pattern of connectivity among the processing nodes. In systems discussed today, this connectivity can be very complex, and in fact the complexity of the connectivity and resulting inter-processor communications is one measure of the processing power of the system. The *Pathfinder* processors are connected in a very simple nearest-neighbor grid, and thus the connectivity of this system is not representative of current neural systems. However, in the application for which the chip was designed, nearest neighbor communication is appropriate and natural. Given the minuscule capability of each processing node in a neural system, the pattern of connectivity imposed on the communications between processors becomes the determining factor in matching the system to the problem to be solved. Although many problems being addressed today require complex inter-processor communication, there will still be examples such as the application addressed by the *Pathfinder* system where a simple pattern of connectivity is the best match for solving the problem.

Like most neural systems under study today, the *Pathfinder* does not operate in isolation, but requires a supportive environment for loading information into the neural system and retrieving results from it. This typically means that the neural system operates as a peripheral unit attached to a more traditional host processing engine. The *Pathfinder* chip operates in exactly this way. The chip's host must first load information into the *Pathfinder* describing the walls in the maze and the starting point for propagation, and then, after the *Pathfinder* does its work, the host must perform the retrace phase of the path-finding algorithm by reading the arrows stored in the chip to determine the required solution to the maze. The work performed by the *Pathfinder* system, propagating information throughout the processing nodes of the network, becomes a single, almost insignificant step as viewed by the host processor that provides the environment for the neural system.

Another characteristic of neural systems displayed by the *Pathfinder* is the notion of a state of activation for each processing node, and an activation rule for modifying the state of each node. In a real neuron, this state of activation is represented by chemical imbalances across the cell membrane. In the *Pathfinder*, the state of activation is represented by the charge on the capacitor in each node. When initialized, each capacitor is fully charged, and remains so until the wavefront of activity originating somewhere in the network reaches it. When the wavefront reaches a given node, the processor at that node becomes actively engaged in the processing activity performed by the network as a whole, and, within that node, the processing activity is indicated by the process of discharging its capacitor, changing the state of activation of that node. When a node's capacitor fully discharges, allowing the wavefront of activity to expand to the next ring of cells, the processing performed by that node is complete, and the state of activation of that node ceases to change. Processing nodes in the neural network that are actively performing useful work can be identified by changes in their state of activation.

By far the most distinguishing quality of neural systems is their implementation of the activation rule by which the state of activation of each processing node is modified. These rules almost always depend upon one or more continuously variable quantities that control how the states of activation of processing nodes change. It is this quality of neural systems, more than any other, that distinguishes these systems from purely digital systems that don't display this dependence on continuously variable, or analog, quantities. One can, for example, design self-sorting memories, and justifiably claim that such a structure is a fine-grained parallel processor, with a processing unit for each word of storage, and a pattern of connectivity between processors which might be as simple as a linear array or something more complex such as a binary tree. However, without the dependence on one or more analog quantities that determine some characteristic of the processing performed, the system is merely a smart memory, or an array processor, and could not be claimed as an example of a neural system.

The *Pathfinder* modifies the state of activation in its processing nodes by discharging the capacitors in those nodes, and the rules for that state modification depend upon three analog quantities that determine the rate at which the capacitor discharges. These three analog quantities, which appeared in Figure 6 as analog voltages on the gates of the transistors at the bottom of the three discharge paths, determine the costs, or weights associated with propagating the frontier of wavefront activity through the processing node in different directions. If one were to write the rule for modifying the state of activation in a given node, it would involve both digital values that specify the direction from which propagation entered the node and whether or not this node is blocked, and it would involve the analog values that control the propagation costs mentioned above. Thus the processing action that occurs at that node is a true hybrid of digital and analog processing. Within the node, no separation of the "digital part" and "analog part" of the circuit is possible. The effects of the two types of variables are tightly intertwined.

Note the correlation between the way in which the *Pathfinder* modifies its state of activation and the way in which a real neuron does so. In a real neuron, signals received from its many inputs are weighted by a set of variable analog quantities and then allowed to proportionally affect the chemical imbalance in the cell of the neuron. This changes the state of activation of the neuron, and when the imbalance reaches some threshold, the neuron "fires" and passes information on to other neurons to which its output is connected. This is very reminiscent of the mousetrap action of the *Mazer* circuit, except that rather than firing immediately when incoming information is received, the neuron delays firing and passing on information for an amount of time that depends on which inputs are received and on what weights are assigned to those inputs. This is a precise description of the *Pathfinder's* operation. The *Pathfinder* delays propagating information for an amount of time that depends on from which direction the wavefront came, and on what weight is assigned for propagation in that particular direction.

Note an important quality demonstrated by both the *Pathfinder* processing nodes and real neurons in passing information throughout the network of interconnected nodes. In both cases, the actual signal transmitted from one node to another is a fully restored digital signal. In the *Pathfinder*, either the capacitor has discharged allowing an arrow latch to flip, or it hasn't. In the neuron, either the cell has "fired" or it hasn't. In no case does a node generate a "partial" signal, or a signal that is only a fraction of its normal value. The transmitted signal is either present or not, in the tradition of digital signals in general. Thus there is no danger of noise accumulating on the data passed from node to node in the network of processors or neurons. At each level of processing, the output generated is fully restored to a valid digital level. However, the time between digital events is a continuously variable quantity, directly affected by the continuous weights associated with modifying the state of activation of the processing node. In the *Pathfinder*, the time delay from the arrival of the wave frontier at a node until the transmission of the frontier to the next node records the cost for traveling through the maze cell represented by that node. In a neuron, the time between firings, or the frequency of the firings of a neuron, records the intensity of the signal being transmitted. In each case, the physical signal being transmitted is digital in nature. The time intervals between signals record the effects of the analog inputs to the computation.

The hybrid nature of the processing that occurs in the *Pathfinder* sets it apart from chip designs that were its contemporaries, and identifies it as an early example of a neural system. The analog variables representing communication weights directly affect the computation that takes place within the processing nodes, despite the fact that the overall problem that this system addresses, maze solving, is an inherently discrete problem involving digital calculations and digital results. By carefully maintaining a fully restored, digital representation for the signals that pass from node to node in the network of processors, so that no information can get lost in accumulated noise, and by using time delays to incorporate analog contributions to the calculations, a successful hybrid system resulted.

LESSONS LEARNED

The design of the *Pathfinder* chip was a unique experience that involved approaches to processor design that had not been tried before. As in any such innovative venture, some lessons were learned from which later designers can benefit. In the case of the *Pathfinder* design, these lessons fall into two categories. Both categories deal with the use of analog variables in neural systems.

The first lesson taught by the *Pathfinder* concerns representation of data. In a system where there are both digital and analog data involved in the processing, some decisions must be made regarding which representation to use for which variables in the processing. Any variable represented in analog form must be tolerant of some inevitable noise on the signal. This noise can be generated by the environment, as for example in capacitive coupling between adjacent wires in the circuit, or it can result from irregularities in the physical medium. An example of the latter is the susceptibility of the *Pathfinder's* cost-setting scheme to variations in threshold voltage of the transistors on whose gates the analog voltages are applied. The design of the circuit must take those sources of noise into account, and minimize their effects. To reduce the problem caused by transistor threshold variation mentioned above, the *Pathfinder* chip used on-chip current mirror circuits so that the actual discharge current levels in the three discharge paths of the processing nodes were set by injecting external known currents into the current mirror inputs, rather than supplying the transistor gate voltages themselves to the chip. The current mirror circuits generated the gate voltages internally, as required by that particular chip, to allow the desired discharge currents to flow, compensating for variations in threshold voltage from chip to chip.

Special attention must be paid to information that passes through many stages of processing, to avoid swamping the data with accumulated noise. Thus, in the *Pathfinder*, the signal that propagates the wavefront of activity from one node to the next is a fully restored digital signal. As a consequence, a given node never misunderstands when the wavefront has reached it. Either the signal has arrived, or it hasn't. Note, however, that the timing of the wavefront's passage does involve analog information. Imagine a straight portion of the wavefront passing as a plane wave from west to east through the array of processors. Certainly, due to irregularities in the processors and the interconnections between them, some of the nodes will pass the wavefront on a little more quickly than others, resulting in slight bulges in what should be a plane wavefront. Fortunately, this affects the operation of the *Pathfinder* system in unimportant ways, resulting in paths that are slightly more costly than minimal being chosen on occasion, but never causing the algorithm to fail due to noise accumulation. This demonstrates a quality that is common to any system that involves analog inputs or parameters. On occasion, results will be generated that are not strictly correct by digital standards, due to noise on the analog inputs or in the analog portion of the processing. This property actually can be used to advantage in neural systems in applications dealing with incomplete data or information that is known only approximately. Remember, people are neural systems, and they sometimes make mistakes too.

The second lesson taught by *Pathfinder* is less mechanical and more thought-provoking than the lesson described above. Testing and analysis of *Pathfinder's* performance demonstrated that it did not assign the analog weights to the proper part of the processing performed by the circuit. The *Pathfinder* assigned weights, or costs, to propagating information through the processing node. The cost was charged for actual travel through the maze cell represented by the node. It was found that assessing costs on travel between cells would have been a more correct choice. Although only subtly different, charging for moving from one cell to the next results in a more accurate reflection of overall path cost than does charging for traversing a given cell. Extended to the more general case of neural systems, this says that the weights should be computed on information where it moves from one processing node to another, rather than computing weights within a given node. More succinctly, costs should be imposed on the communication, and not the computation, that occurs in neural systems. Most discussions of neural systems today include this notion, though at the time of the *Pathfinder*, this was a novel idea.

In summary, the *Pathfinder* system for maze solving is an example of a system that demonstrates many of the characteristics associated with neural systems today. Its fine-grained parallel architecture and its system of analog weights imposed on inter-processor communication mark it as an early example of this class of circuits. The techniques used in its design, and the lessons learned from using the system may benefit current designers of circuits using similar approaches to this hybrid combination of digital and analog circuitry.

ACKNOWLEDGEMENTS

The author wishes to acknowledge the guidance and insights provided by C.A. Mead and I.E. Sutherland during their supervision of the research described in this paper.

REFERENCES

1. Akers, S., "A Modification of Lee's Path Connection Algorithm," *IEEE Transactions on Electronic Computers* (Short Notes), Vol. EC-16, pp. 97-98, February, 1967.

2. Carroll, C.R., "A Smart Memory Array Processor for Two Layer Path Finding," *Proceedings of the Second Caltech Conference on Very Large Scale Integration*, Caltech Computer Science Department, pp. 165-195, 1981.

3. Carroll, C.R., "Hybrid Processing," Ph.D. Thesis, Computer Science Department, California Institute of Technology, 1982.

4. Lee, C., "An Algorithm for Path Connections and its Applications," *IEEE Transactions on Electronic Computers* , Vol. EC-10, pp. 346-365, September, 1961.

5. Moore, E., "Shortest Path Through a Maze," *Annals of the Computation Laboratory of Harvard University*, Vol. 30, Cambridge, MA: Harvard University Press, pp. 285-292, 1959.

6. Sutherland, I.E., "A Better Mousetrap," Computer Science Department display file #562, California Institute of Technology, March 8, 1977.

Christopher R. Carroll *was born in Cincinnati, Ohio on February 27, 1954. He received the Bachelor of Engineering Science degree from Georgia Institute of Technology in June, 1975, and the Master of Science in Electrical Engineering and Ph.D. in Computer Science degrees from California Institute of Technology in June, 1977 and June, 1982, respectively. His Ph.D. research involved the design and development of the two integrated circuits described in this paper, under the direction of C. A. Mead and I. E. Sutherland. After serving as Assistant Professor of Electrical Engineering at Duke University from 1981 until 1988, he is now Associate Professor of Computer Engineering at the University of Minnesota, Duluth. His research interests include special purpose digital systems, VLSI design, and microprocessor applications, especially as they relate to educational environments.*

2

RESISTIVE FUSES: ANALOG HARDWARE FOR DETECTING DISCONTINUITIES IN EARLY VISION

John Harris, Christof Koch, Jin Luo
Computation and Neural Systems Program
California Institute of Technology
Pasadena, California 91125

and

John Wyatt
Department of Electrical Engineering and Computer Science
and Research Laboratory of Electronics
Massachusetts Institute of Technology
Cambridge, Massachusetts 02139

Abstract: The detection of discontinuities in motion, intensity, color, and depth is a well studied but difficult problem in computer vision. We discuss our "resistive fuse" circuit—the first hardware circuit that explicitly implements either analog or binary line processes in a controlled fashion. We have successfully designed and tested an analog CMOS VLSI circuit that contains a 1-D resistive network of fuses implementing piece-wise smooth surface interpolation. The segmentation ability of this network is demonstrated for a noisy step-edge input.

We derive the specific current-voltage relationship of the resistive fuse from a number of computational considerations, closely related to the early vision algorithms of Koch, Marroquin and Yuille (1986) and Blake and Zisserman (1987). We discuss the circuit implementation and the performance of the chip. In the last section, we show that a model of our resistive network—in which the resistive fuses have no internal dynamics—has an associated Lyapunov function, the co-content. The network will thus converge, without oscillations, to a stable solution, even in the presence of arbitrary parasitic capacitances throughout the network.

INTRODUCTION

Most early vision algorithms incorporate the generic constraint that variables such as surface orientation and reflectance, depth or optical flow vary slowly in space (Marr and Poggio, 1976; Grimson, 1981; Ikeuchi and Horn, 1981; Horn and Schunck, 1981; Terzopoulos, 1983; Hildreth, 1984; Poggio, Voorhees and Yuille, 1985; Nagel, 1987). Within the standard regularization approach, this is reflected in the use of stabilizing operators corresponding to various measures of smoothness (Poggio, Torre and Koch, 1985). Thus, in the problem of interpolating a 2-D surface through sparse and noisy depth measurement, the final surface should be as close as possible to the initial data as well as being as smooth as possible (Grimson, 1981); or, in the problem of computing optical flow from the time-varying intensity, the final flow field should be compatible with the locally measured velocity data as well as being smooth (Horn and Schunck, 1981; Hildreth, 1984; Nagel, 1987). However, surfaces display discontinuities where the smoothness constraint is violated. Thus, the to-be-reconstructed surface may have been generated by an underlying piece-wise smooth or even piece-wise constant depth distribution. Or, the 2-D velocity field induced by a rigid object moving/rotating in an otherwise stationary environment varies smoothly across the surface of the object but is zero beyond the contours of the object (since the background is stationary).

In the last years, a number of researchers have introduced powerful algorithms to deal with the representation of such discontinuities. Geman and Geman (1984) first proposed binary line processes to model discontinuities in intensity within the stochastic framework of Markov Random Fields. Discontinuities are subject to various constraints, such that they should form along continuous contours, should not intersect nor form parallel lines. Their approach was extended and modified to account for discontinuities in depth, texture and color by Poggio and his collaborators (Marroquin, Mitter and Poggio, 1984; Poggio, Gamble and Little, 1988) as well as to discontinuities in the optical flow (Hutchinson, Koch, Luo and Mead, 1988). The principal drawback of the Geman and Geman-type method is the computational expense involved in minimizing the associated non-convex cost functionals using stochastic optimization methods, in particular when numerous constraints (e.g. continuity of discontinuities) are incorporated. A number of authors have used deterministic methods to find the (local) minimum of the associated convex or non-convex variational functionals, with next-to-optimal results (Terzopoulos, 1986; Koch, Marroquin and Yuille, 1986). A rigorous deterministic approach has been championed by Blake and Zisserman (1987). Their "graduated non-convexity" (GNC) algorithm bears many similarities to the above methods, and leads to excellent results in the case of piece-wise continuous reconstruction of surfaces (Blake, 1989).

Poggio and Koch (1985) show how standard regularization algorithms can map onto simple resistive networks. Finding the minimum of the standard regularized and quadratic cost functional is equivalent to finding the state of least power dissipation in an appropriate electrical network, where the data are given by injecting current into certain nodes and the solution by the stationary voltage distribution. Figure 1 shows the appropriate network for membrane-type surface interpolation, where the "strength" of smoothing is given by the value of the horizontal grid conductance. For an overview of analog circuits for implementing early vision algorithms see Koch (1989) and Horn (1989).

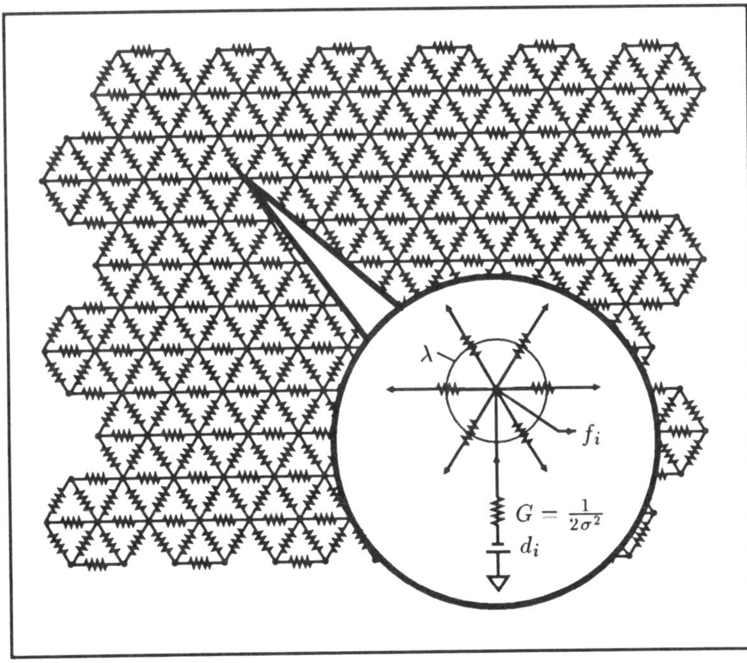

Figure 1 Resistive network for fitting the smoothest surface f through sparse and noisy data d. The circuit minimizes the variational functional of the two-dimensional extension of eq. (1) in the absence of line discontinuities. In the continuum limit, minimization of this functional corresponds to the Euler-Lagrange equation $\lambda\nabla^2 f + Gf = Gd$. The battery supplies the measured depth data d_i, while the vertical conductance G corresponds to $1/(2\sigma^2)$ and the horizontal conductance of the grid to λ. If no data are present at a particular location i, G is set to zero. The stationary voltage distribution then corresponds to the interpolated surface f_i. The amplitude of the horizontal grid conductance, λ, controls the amount of smoothing. A 48 by 48 pixel hexagonal network has been built and tested successfully (Luo, Koch and Mead, 1988).

The recent development of subthreshold, analog CMOS VLSI circuits for various sensory tasks by Carver Mead (see in particular his recent textbook, Mead, 1989) has enabled us to implement these resistive networks—together with the photo-transduction stage—using this real-time, low power and robust technology. Two circuits are particularly attractive for our purposes: a photo-transistor with a logarithmic voltage output over five orders of intensity brightness (Mead, 1985, 1989) and a transistor circuit with a linear current-voltage relationship for small voltage gradients (Sivilotti, Mahowald and Mead, 1987; Mead, 1989). The value of the slope, i.e. the resistance, can be varied over five orders of magnitude. Using this as our basic construction element, we built and tested a 48 by 48 pixel resistive network for smoothing and interpolating noisy and sparse data (Luo, Koch and Mead, 1988; see Fig. 1).

We introduce in this paper an analog, purely deterministic approach to locating discontinuities in the case of interpolating noisy and sparsely sampled depth data. It leads to a very simple and elegant circuit implementation in terms of a two-terminal, nonlinear, voltage-controlled resistor termed "resistive fuse" (Harris and Koch, 1989). We have implemented this device in analog CMOS and demonstrate its performance here.

THEORY

Let us begin by justifying "resistive fuses" as specialized circuit elements for implementing discontinuities. Since our methodology does not distinguish between a 1-D and a 2-D implementation of smoothing in the presence of discontinuities, we will first consider the 1-D case. The simplest possible variational functional for interpolating noisy and sparsely sampled data d_i in the presence of binary line discontinuities ℓ_i is a membrane type of surface interpolation:

$$J(f,\ell) = \lambda \sum_i (f_i - f_{i+1})^2(1-\ell_i) + \frac{1}{2\sigma^2}\sum_i (d_i - f_i)^2 + \alpha \sum_i \ell_i, \quad (1)$$

where f_i is the value of the final surface f at location i, σ^2 the variance of the additive Gaussian noise process assumed to corrupt the data d_i and λ and α are free parameters. The first term in this functional implements the constraint that surfaces should, in general, vary smoothly. If all variables, with the exception of f_i, f_{i+1} and ℓ_i, in eq. (1) were held fixed and $\lambda(f_i - f_{i+1})^2 < \alpha$, it would be "cheaper" to pay the price $\lambda(f_i - f_{i+1})^2$ and set $\ell_i = 0$ than to pay the larger price α. However, if the gradient becomes too steep, the line process is switched on, i.e. $\ell_i = 1$, and the "price" α is paid. The second term in eq. (1), where the sum only includes those locations i where data exist, forces the final solution f to be close to the measured data d. How close depends on the estimated magnitude of the noise, in this case on σ^2. Thus, the surface f, with its associated set of discontinuities ℓ, minimizing eq. (1) will be the one that best satisfies the conflicting demands of piece-wise smoothness and fidelity to the measured data. The functional of eq. (1) is non-convex and a large number of both stochastic and deterministic methods have been designed to find optimal or nearly optimal solutions for this and similar functionals (Geman and Geman, 1984; Marroquin, Mitter and Poggio, 1987; Koch, Marroquin and Yuille, 1986; Blake and Zisserman, 1987; Terzopoulos, 1983, 1986).

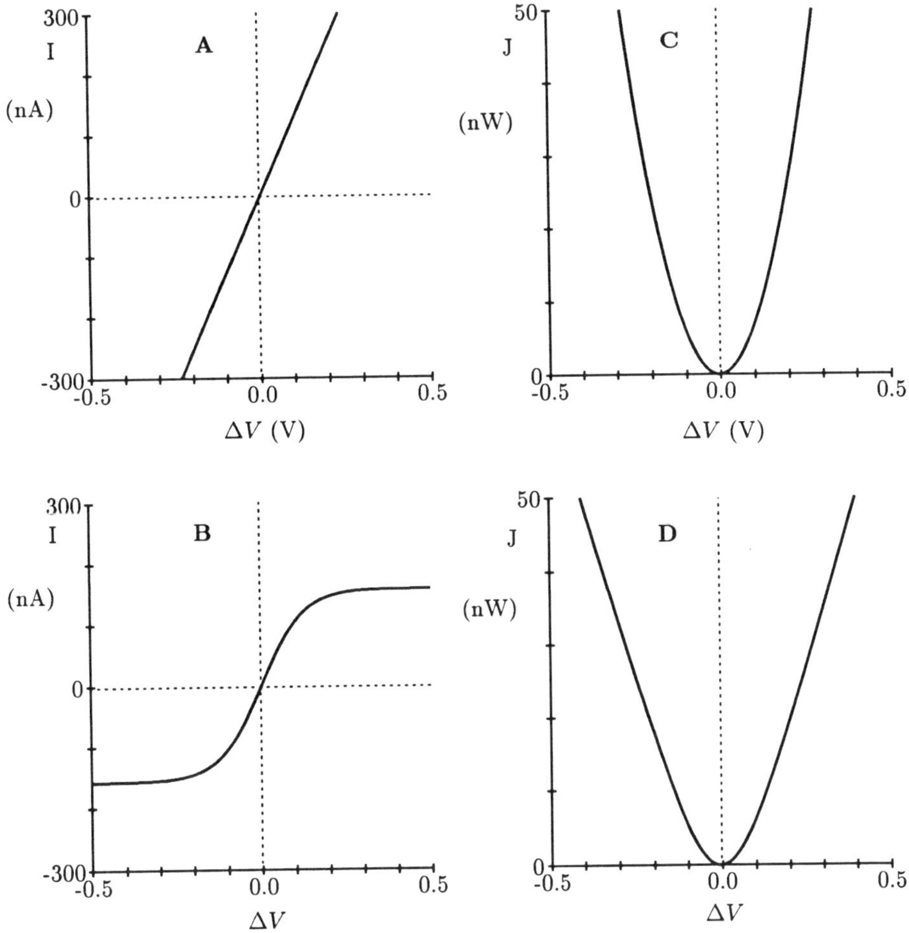

Figure 2 Theoretical I-V curves for a linear resistor (A) and a measured I-V curve for Mead's saturating resistor (B). Integrating numerically over these curves gives the co-content of the linear resistor (C) and the saturating resistor (D). Co-content is defined by eq. (2) and represents generalized power for non-linear systems. The co-content for the linear resistor is equivalent to half the dissipated power, and thus a quadratic function in ΔV, while the co-content for the saturating resistor becomes a linear function of ΔV as $|\Delta V| \rightarrow \infty$.

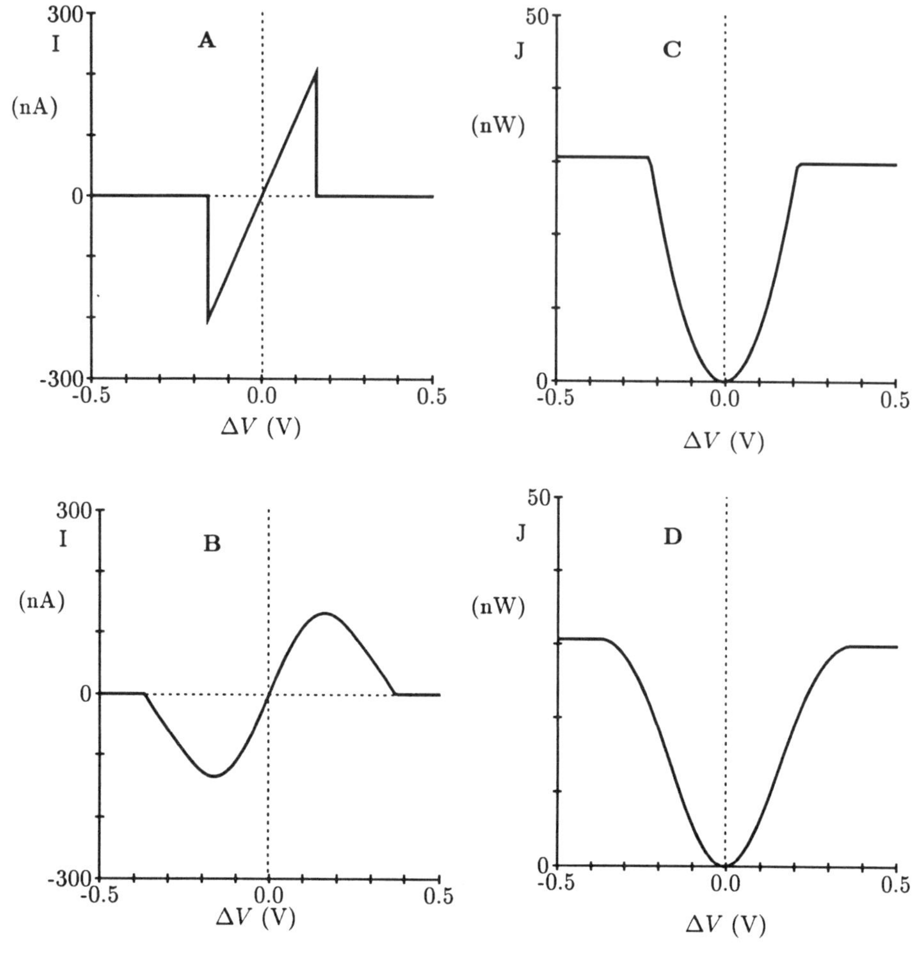

Figure 3 Theoretical I-V curve for an infinite-gain fuse (A) and a measured I-V curve for a finite-gain resistive fuse (B). Integrating numerically over these curves gives the co-content J for the infinite-gain (C) and the finite-gain fuse (D).

Figure 3C shows a plot of $J(f, \ell)$ as a function of the depth at locations f_i and f_{i+1} and as a function of the discontinuity ℓ_i. The values of the surface and of the line discontinuities are assumed to be fixed at all other locations. As long as $\lambda(f_i - f_{i+1})^2 \leq \alpha$, the function E is quadratic in the gradient. However, once $|f_i - f_{i+1}|$ exceeds the gradient limit $\sqrt{\alpha/\lambda}$, E remains flat at $E = \alpha$, independent of the magnitude of $f_i - f_{i+1}$ (Blake and Zisserman, 1987).

The appropriate circuit implementation is a straightforward modification of the network shown in Fig. 1. The surface f_i represents the final reconstructed points. The voltage on the battery is d_i, and the conductance G equals $1/(2\sigma^2)$. If no measured surface value d is present at a particular location, $G = 0$ at that location. The value of the grid conductance λ controls the amount of smoothing. Binary switches, breaking the resistive connections among neighboring nodes, would implement discontinuities in the surface. As long as the switch is closed, the current is linear in the voltage drop across the device. Since the electrical power in a linear network is proportional to the square of the voltage gradient across all resistances, the power is quadratic in the gradient and can thus be identified with the first term in eq. (1). Once the threshold has been exceeded, the binary switch opens and no more current flows through the device. The digital processors controlling the switches need access to the state of the neighboring switches as well as to the neighboring depth values. We will now demonstrate, however, how this mixed analog-digital circuit can be replaced by a single analog non-linear resistor, the "resistive fuse."

The circuit implementation of binary discontinuities will require nonlinear circuit components. As pointed out by Poggio and Koch (1985), the notion of minimizing power in linear networks implementing quadratic "regularized" algorithms must be replaced by the more general notion of minimizing the total resistor co-content (Millar, 1951). For a two-terminal voltage-controlled resistor characterized by $I = f(V)$, the co-content is defined as

$$J(V) = \int_0^V f(V')dV'. \tag{2}$$

For a linear resistor, $I = GV$, the co-content is given by $\frac{1}{2}GV^2$, which is just half the dissipated power $P = GV^2$ (Fig. 2). For a network consisting of a collection of resistors, voltage sources and other elements, the total network co-content is defined as the sum of all the (linear or nonlinear) resistor co-contents, that is,

$$J_{total}(t) = \sum_{\text{all resistors}} J_k(V_k(t)). \tag{3}$$

The co-content for various resistors is plotted in Figs. 2 and 3. Differentiating eq. (2), we have:

$$f(V) = \frac{dJ}{dV}. \tag{4}$$

The appropriate current-voltage relationship of an *infinite-gain resistive fuse* is illustrated in Fig. 3A. As long as the voltage drop across this device is below the threshold, the current through the nonlinear resistor is linearly related to the voltage across it. Once past the threshold, the circuit breaks (hence the name "fuse"), and the current is zero for all values of the voltage gradient. This two-terminal device then implements the high-level constraint that surfaces should be smooth unless their neighboring values differ by more than $\pm\sqrt{\alpha/\lambda}$, at which point the surfaces will break.

The I-V relationship of the device we have built is shown in Fig. 3B. The most salient difference from the infinite-gain fuse are the smooth flanks, where the current decreases smoothly to zero for increasing values of the voltage gradient [1]. (in contrast with the discontinuity in the I-V relationship for the infinite-gain fuse). In this region the slope conductance dI/dV will be negative (Fig. 13C) Our measured I-V curve can be related directly to the concept of analog line discontinuities of Koch *et al.* (1986). The key idea is that, following Hopfield and Tank (1985) in their neural network implementation of the Traveling Salesman Problem, binary discontinuities are mapped onto continuous "neurons," whose output is constrained to lie between 0 and 1. The input-output relationship of these "discontinuity neurons" is governed by the sigmoidal function $V = g(U)$, where $g(U)$ is a strictly monotonic function, usually taken to be

$$g(U) = \frac{1}{1 + e^{-2\eta U}}, \tag{5}$$

with the "gain" $\eta > 0$. The network converges to a stationary solution using a steepest descent rule. The solutions obtained were qualitatively very similar to the solutions obtained with binary line processes. It is rather straightforward to derive an "analog" version of resistive fuses (Harris, Koch, Staats, Luo and Wyatt, 1989), with the following I-V relationship

$$I = f(V) = \left[1 - g(\frac{V^2 - \alpha}{\beta})\right] V, \tag{6}$$

where $\beta > 0$ is a parameter related to the analog line process implementation (identical to c_G of eq. (7c) of Koch *et al.*, 1986). Our measured I-V curve for the fuse (Fig. 3B) implements this function. For $\eta \to \infty$, the function g becomes binary and $f(V)$ of eq. (6) approaches the form of the infinite-gain fuse (Fig. 3A).

So far we have only discussed the implementation of binary or analog discontinuities in 1-D. For 2-D image problems, horizontal as well as vertical line

---

[1] The I-V characteristic of our experimental fuse relates somewhat to the theoretical work of Perona and Malik (1988) who simulated a network of elements with similar I-V characteristics to perform image segmentation.

processes need to be incorporated into the variational functional. Furthermore, it has been standard practice to constrain the geometry of line processes by adding appropriate terms to the 2-D extension of eq. (1). Some of the more common constraints are that discontinuities should occur along continuous contours, should not intersect nor form along parallel lines (Geman and Geman, 1984). Furthermore, Poggio *et al.* (1988) introduced the notion that discontinuities in depth should in general coincide with discontinuities in intensity, that is intensity edges.

We previously demonstrated how a piece-wise smooth optical flow field, induced by moving objects, can be successfully recovered in the presence of binary motion discontinuities with the above set of constraints (Koch *et al.*, 1986; Hutchinson *et al.*, 1988). We repeated these simulations using only the finite-gain resistive fuses of eq. (6) together with the constraint that motion discontinuities should only occur together with intensity discontinuities, in our case zero-crossings of the $\nabla^2 G$ operator. The performance of both algorithms—for 128 by 128 video image sequences of several moving and partially occluding people—is very similar (for more details see Harris *et al.*, 1989). Since the co-localization of all or most motion discontinuities with intensity discontinuities (but not necessarily the reverse) is relatively simple to implement at the circuit level, we feel that we can now design VLSI circuits to compute intensity, motion and depth discontinuities for real, two-dimensional images. The following section discusses the detailed circuit implementation of the resistive fuse.

CIRCUIT DETAILS

The circuit schematic for the fuse is shown in Fig. 4. The circuitry above the dotted line in the figure is Mead's saturating resistor (Mead, 1989) with a p-type pullup transistor that sets the nominal resistance of the fuse. In subthreshold operation, the current through a transistor varies exponentially with the gate-to-source voltage. Thus, the voltage V_B produces a current I_B equal to:

$$I_B = I_0 e^{\kappa(V_{DD} - V_B)} \tag{7}$$

Following Mead (1989), all voltages are assumed to be normalized by kT/q. The variable κ is a process-dependent parameter that reflects the inability of the gate to be 100% effective in reducing the barrier potential. I_0 is a constant that includes the width and length of the transistor as well as process-dependent fabrication parameters. Letting $I_F = I_B$, the I-V relation of the resistor can be derived as:

$$I_{FUSE} = \frac{I_F}{2} \tanh\left(\frac{\Delta V}{2}\right) \tag{8}$$

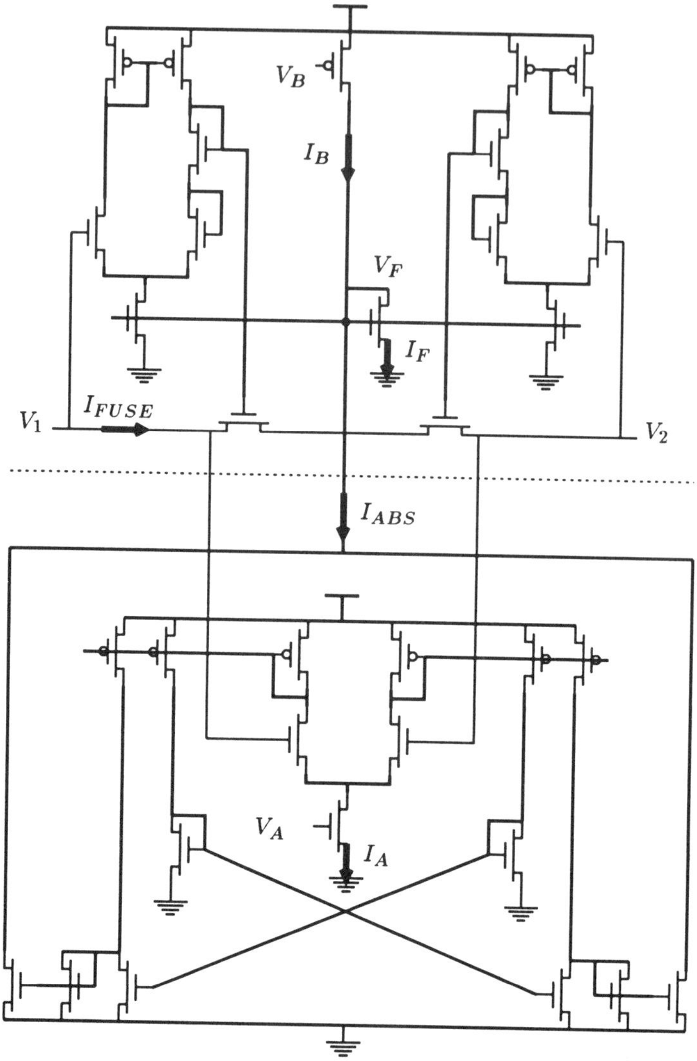

Figure 4 Schematic of the fuse circuit. The nonlinear, voltage-controlled resistance is seen across the V_1 and V_2 terminals. The circuitry above the dotted line is a saturating resistor (Mead, 1989) with V_B controlling the nominal amount of resistance. The circuit below the dotted line is a saturating absolute-value circuit that turns off the resistor for large $|V_1 - V_2|$. V_A determines the magnitude of the current pulled away by the absolute-value circuit.

where $\Delta V = V_1 - V_2$. For small ΔV this portion of the circuit operates as a linear resistor with a resistance of

$$R = \frac{4kT/q}{I_F} \tag{9}$$

Because we are working in the subthreshold region, I_F and thus the resistance can be varied over five orders of magnitude. For large ΔV the resistor saturates and provides a constant current of $I_F/2$. A measured I-V curve for this circuit is shown in Fig. 2B.

The circuit below the dotted line in the figure performs a saturating absolute-value operation. This portion of the circuit is enabled by the voltage V_A, which creates a current I_A equal to:

$$I_A = I_0 e^{\kappa V_A} \tag{10}$$

The positive parts of the outputs of a dual-output wide-range transconductance amplifier are combined to create a current of:

$$I_{ABS} = I_A \tanh\left(\frac{\kappa|\Delta V|}{2}\right) \tag{11}$$

By Kirchhoff's current law, the current I_F is:

$$I_F = \lfloor I_B - I_{ABS} \rfloor \tag{12}$$

where the symbols $\lfloor\ \rfloor$ are defined as

$$\begin{aligned} \lfloor x \rfloor &= x \quad \text{if} \quad x \geq 0 \\ &= 0 \quad \text{if} \quad x < 0 \end{aligned}$$

Substituting (11) and (12) into eq. (8), gives

$$I_{FUSE} = \frac{1}{2}\left\lfloor I_B - I_A \tanh\left(\frac{\kappa|\Delta V|}{2}\right)\right\rfloor \tanh\left(\frac{\Delta V}{2}\right) \tag{13}$$

When $|\Delta V|$ is small, the fuse acts as a linear resistor whose nominal resistance is set by I_B. When $|\Delta V|$ is large, I_A increases above the current supplied by the p-type pull-up, and V_F is pulled to ground, shutting off the resistor. In between these extremes, the fuse exhibits a gradual transition.

Figure 5 shows a family of curves measured by varying V_A while keeping V_B constant. By varying V_A in this way, the circuit's I-V characteristic can be continuously and smoothly changed from that of a saturating resistor to

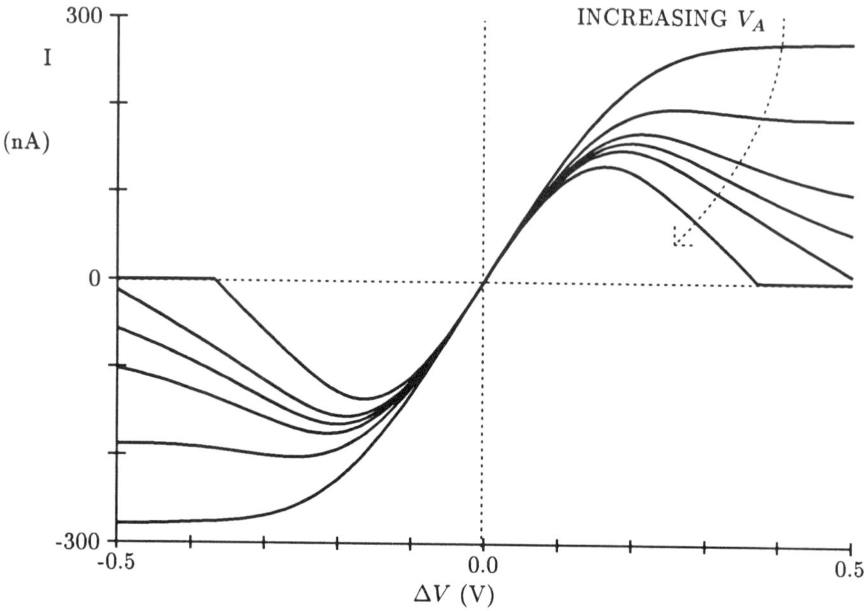

Figure 5 Measured I-V curves that show the effect of continuously varying from the saturating characteristic to that of the fuse curve. V_B was set to 4V and V_A was varied from 0V to 2V. When $V_A = 0$, the resulting I-V curve is identical to that of Mead's saturating resistor.

the fuse I-V curve. Setting $V_A = 0$ gives $I_A = 0$ disabling the absolute-value circuit, and giving the fuse a saturating I-V relationship (Fig. 2B).

Integration of the I-V curves in Fig. 5 gives the family of co-content curves shown in Fig. 6. For small ΔV the co-content is quadratic and for large ΔV the co-content saturates at a constant value. Instead of saturating for large voltage differences, the co-content of the saturating resistor increases linearly with voltage. As will be seen in the following section, networks of resistors with positively sloped I-V curves are guaranteed to converge to a single unique minimum value of the co-content. By turning the voltage control, we are changing the energy landscape in a continuous fashion ("continuation method") from containing one unique global minimum to a landscape containing many local minima.

The fuse provides a mechanism for changing the threshold value. If we assume that the circuit is operating in the linear region of the two hyperbolic tangents, I_{FUSE} becomes twin parabolas of the form:

$$I_{FUSE} = \left[\frac{I_B}{4} - \kappa \frac{I_A}{8} |\Delta V| \right] \Delta V \tag{14}$$

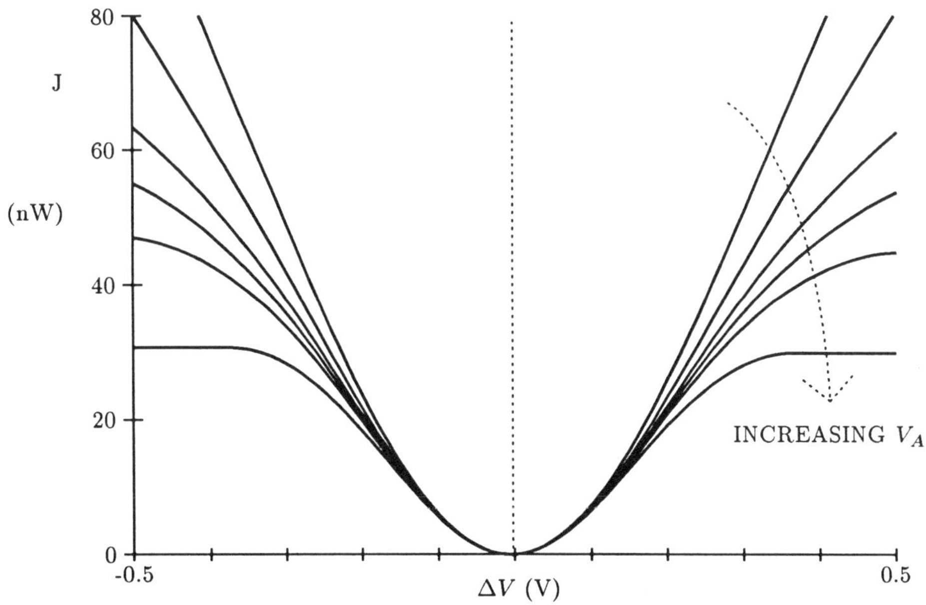

Figure 6 Co-content functions: each curve was numerically integrated from the family of curves in Fig. 5. Continuously varying the co-content curves in this way performs a useful computation that is explored more in Fig. 10 and Fig. 11.

This linear analysis indicates that the measured curve in Fig. 3B consists of a parabola in each of the first and third quadrants. This current in eq. (14) is cut to zero for:

$$|\Delta V| \geq 2\frac{I_B}{I_A}\frac{kT}{q\kappa} \tag{15}$$

I_{FUSE} reaches extremum points at:

$$|\Delta V| = \frac{I_B}{I_A}\frac{kT}{q\kappa} \tag{16}$$

The extremum points can be set by the ratio of I_B to I_A. In subthreshold operation, the width of the saturating tanh curves is about 100mV. The extremum points can then only be be varied from 0 to about $\pm$100mV. For gate voltages above the threshold of the bias transistors, the width of the linear region of the hyperbolic tangent function increases by $V_{GS} - V_T$, where V_{GS} is the gate-to-source voltage and V_T is the threshold voltage of the bias transistors. Thus, by going slightly above threshold the extremum point can be varied from 0 to

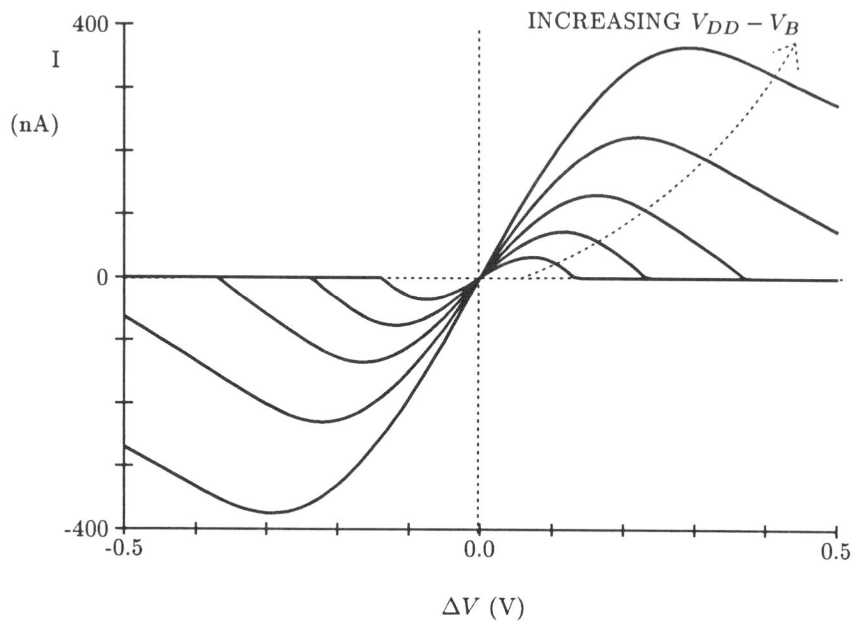

Figure 7 Measured I-V curves illustrating different line process penalties. V_A was kept constant at 2V and V_B was varied from 3.9V to 4.1V.

about ±500mV. Figure 7 shows a family of I-V curves measured by varying V_B and holding V_A constant.

We are studying the use of a high-gain fuse, a circuit that does not have a large incrementally active region in its I-V curve (Fig. 8). Circuit simulations of the high-gain fuse show I-V curves that look like those of the infinite-gain fuse in Fig. 3A. Instead of feeding the absolute-value current back to the resistor bias circuits, current is fed back to a pass gate that acts as a binary switch in the current path. When $I_B > I_{ABS}$ the voltage on the gate of the binary switch (V_F) is charged to V_{DD}. On the other hand, when $I_B < I_{ABS}$, V_F is pulled to ground, effectively open-circuiting the resistor. The resistance of the resistor is controlled by V_R, which sets the bias current I_R. Notice that the current that controls the line process penalty is decoupled from the current that sets the resistance of the fuse. Assuming high-gain elements, the I-V equation for the high-gain fuse is given by:

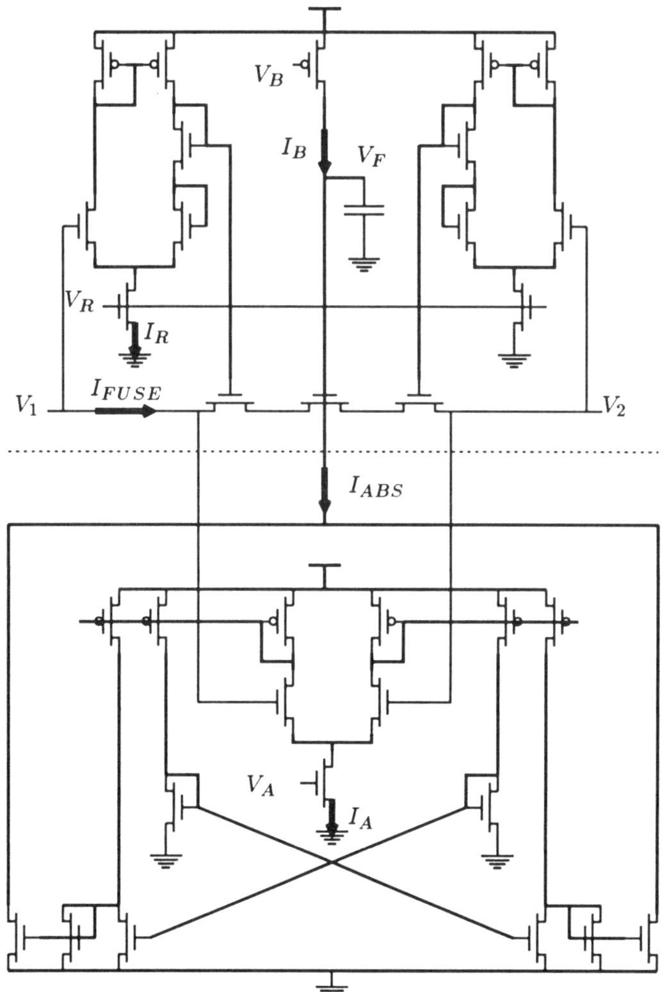

Figure 8 Modification of the fuse to obtain a high-gain characteristic. As before, a saturating resistor and an absolute-value circuit are combined to create a fuse. However, different from the circuit of Fig. 4, the absolute-value circuit discharges the gate of a pass transistor that has been added in the resistance path. This pass gate acts as a binary switch that is opened or closed dependent on whether or not the absolute-value current is greater than the threshold current provided by V_B. V_R provides independent control of the resistance of the fuse when the binary switch is closed.

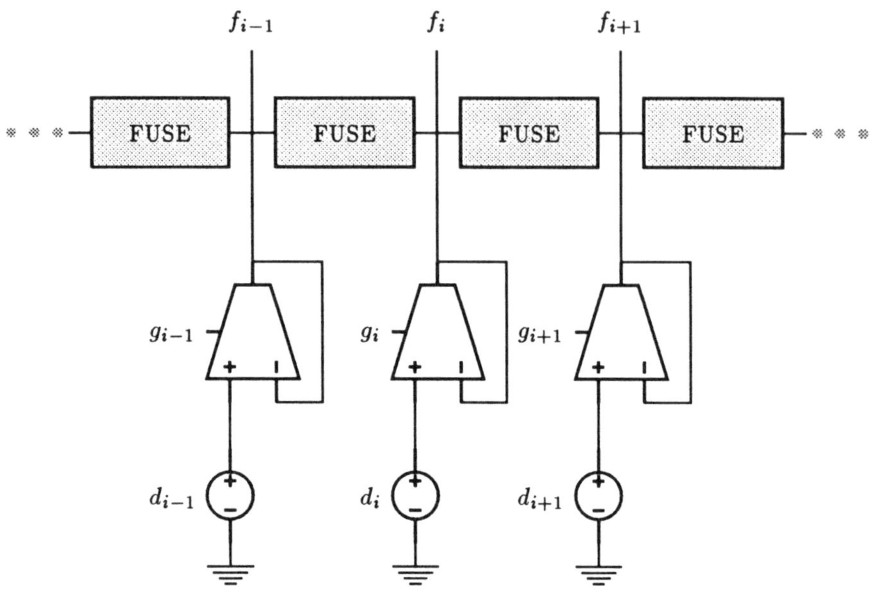

Figure 9 Layout of the 1-D fuse network. Voltage sources d_i provide input to the network through wide-range transconductance amplifiers. The bias voltages on these amplifiers g_i controls their conductance. The smoothed and segmented outputs are given as voltages at f_i. This network was designed to implement eq. (1).

$$\begin{aligned} &\text{if } \; I_A \tanh\left(\frac{\kappa|\Delta V|}{2}\right) < I_B \text{ then } I_{FUSE} = \frac{I_R}{2}\tanh\left(\frac{\Delta V}{2}\right) \\ &\text{if } \; I_A \tanh\left(\frac{\kappa|\Delta V|}{2}\right) > I_B \text{ then } I_{FUSE} = 0 \end{aligned} \tag{17}$$

This implementation of the fuse shares an advantage with Mead's saturating resistor layout, because only one biasing circuit is needed for each node. This saves many transistors, especially in 2-D layouts. The low-gain fuse requires 33 transistors per connection, while the high-gain fuse requires only 21 transistors per connection plus 6 transistors per node. For a hexagonal mesh, each basic cell needs to contain one node plus half of the six neighboring connections, requiring a total of 69 transistors per cell for the high-gain fuse and 99 transistors per cell for the low-gain version.

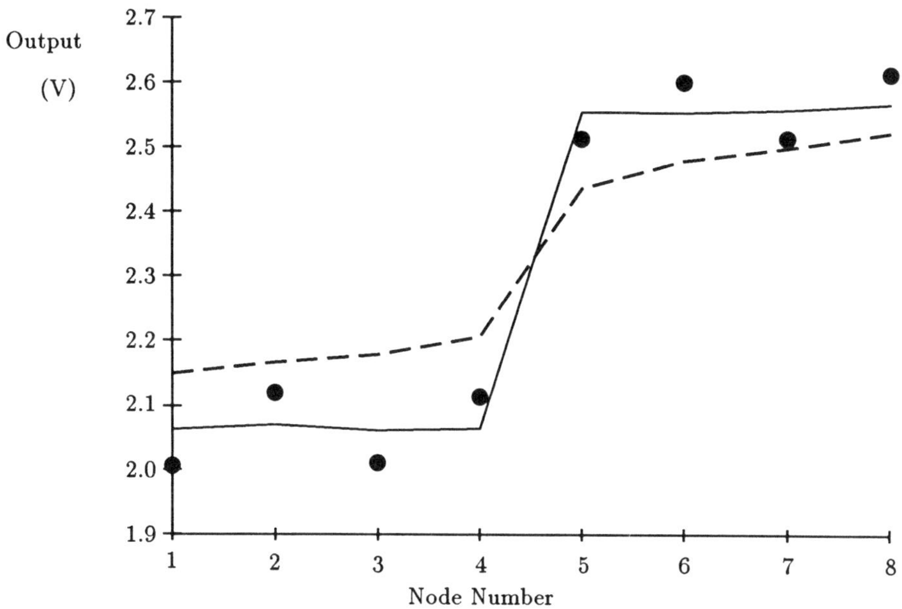

Figure 10 Measured segmentation from an experimental resistive fuse network. The circles denote "noisy" step data that was used as the input to the network. The solid-line curve indicates measured voltages from the chip. The dotted-line curve shows the measured voltage output given by a network of Mead's saturating resistors.

A network of eight fuses (of the type shown in Fig. 4) was fabricated and successfully demonstrated. The schematic is shown in Fig. 9. Eight voltage values are input as the d_i values. The smoothed and segmented f_i voltages are the resulting outputs. Figure 10 shows a segmentation result for a "noisy" 1-D step edge. The network effectively smooths out small steps without degrading large step edges. The I-V curves of the fuses in this example have been set to the form shown in Fig. 3B. In this configuration, the network exhibits a hysteresis property in which two stable final states are possible. The two stable states correspond to segmenting or smoothing the step edge. The segmented stable state is shown as the solid line in Fig. 10. The smoothed stable state becomes essentially a flat horizontal line. The final state depends on the temporal history of the network. To ensure that the proper stable state is reached in a deterministic fashion, V_A is initially set to 0V and then gradually moved to its final value.

The hysteresis properties of the network can be better understood through a load-line analysis of a much simplified circuit (Fig. 11). The current through

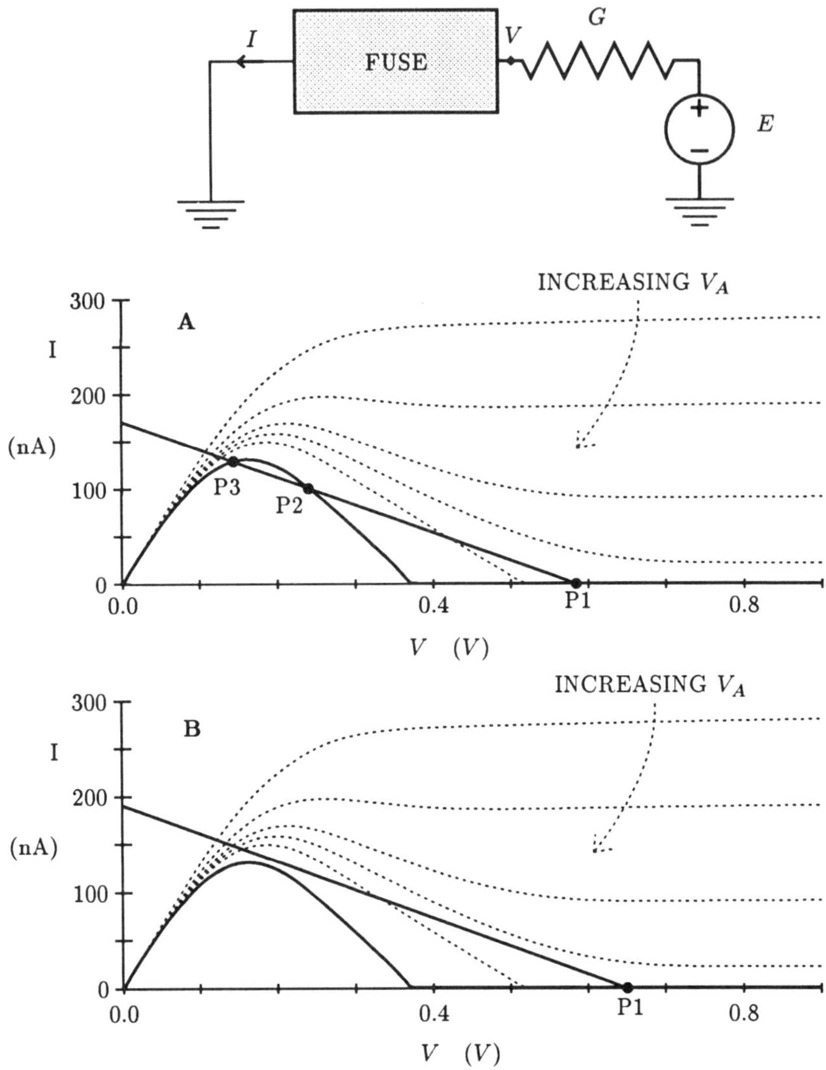

Figure 11 Simple load-line analysis shows that there can be up to three equilibrium points for the fuse/resistor circuit given above. The I-V curves for the measured fuse and the simulated voltage source/resistor are shown as solid lines. For plot A, points P1 and P3 are stable, and P2 is unstable. Voltages in the neighborhood of P2 will be driven to either P1 or P3. By increasing the value of the voltage source E, a single stable equilibrium point P1 remains (plot B). The dotted-line curves show the effect of changing V_A.

the fuse is plotted as a function of the voltage across the fuse. The simulated voltage source/resistor is also illustrated as a solid line, with the negative slope of this line given by the conductance G and the x-intercept given by the value of the voltage source E. A stability analysis reveals that the system possesses up to three equilibria. In the case illustrated in Fig. 11A, the middle equilibrium is unstable and the voltage will tend toward the two stable solutions P1 and P2. Point P1 corresponds to segmentation, and P3 corresponds to smoothing. By increasing the value of the voltage source E (Fig. 11B), only a single stable equilibrium point remains, corresponding to segmentation. Of course, stability cannot be guaranteed for negative values of G. The dotted-line curves show the effect of changing V_A.

Figure 12 shows the computed total co-content from the I-V curves shown in Fig. 11. For Fig. 12A, P1 is the global and P3 is only a local minimum, while P2 corresponds to an unstable local maximum. In contrast, Fig. 12B contains a single equilibrium point, P1, which corresponds to a discontinuity. The dotted lines show the effect of increasing V_A, deforming the energy surface from one with a single equilibrium point to one with two local minima. By using a continuation method in this fashion, discontinuities are deterministically located. Reasonable performance may be obtained by using a single setting of the fuse control voltages and keeping the voltages constant over time. This static approximation of the continuation method will still smooth small step edges while preserving large steps. However, medium steps, such as those simulated in Fig. 11, can be either smoothed or segmented depending upon the temporal history of the network. This load-line analysis is a simplified version of the true dynamics of networks of fuse elements, but serves to illustrate the complexity of even a single fuse element circuit.

STABILITY

Though the chord resistance of the fuse circuit is always positive, its incrementally negative resistance regions (see Fig. 13) raise doubts about the stability of networks of resistive fuse elements. One question that has already been alluded to above is the issue of whether the network will converge at all and whether a unique stationary solution exists. The reasoning presented later in this section supports the following conclusions.

1. Monotonic Resistors

Suppose all the nonlinear resistors are *incrementally strictly passive*, i.e., have I-V curves with positive slope, $dI/dV > 0$, everywhere. One instance of such a device is Mead's saturating resistor (Fig. 2B). Then the stationary network solution for a given input image will be unique. If we further suppose that the nonlinear resistors are ideal memoryless elements (i.e., that we can neglect the fast parasitic dynamics internal to each resistor circuit), then the network

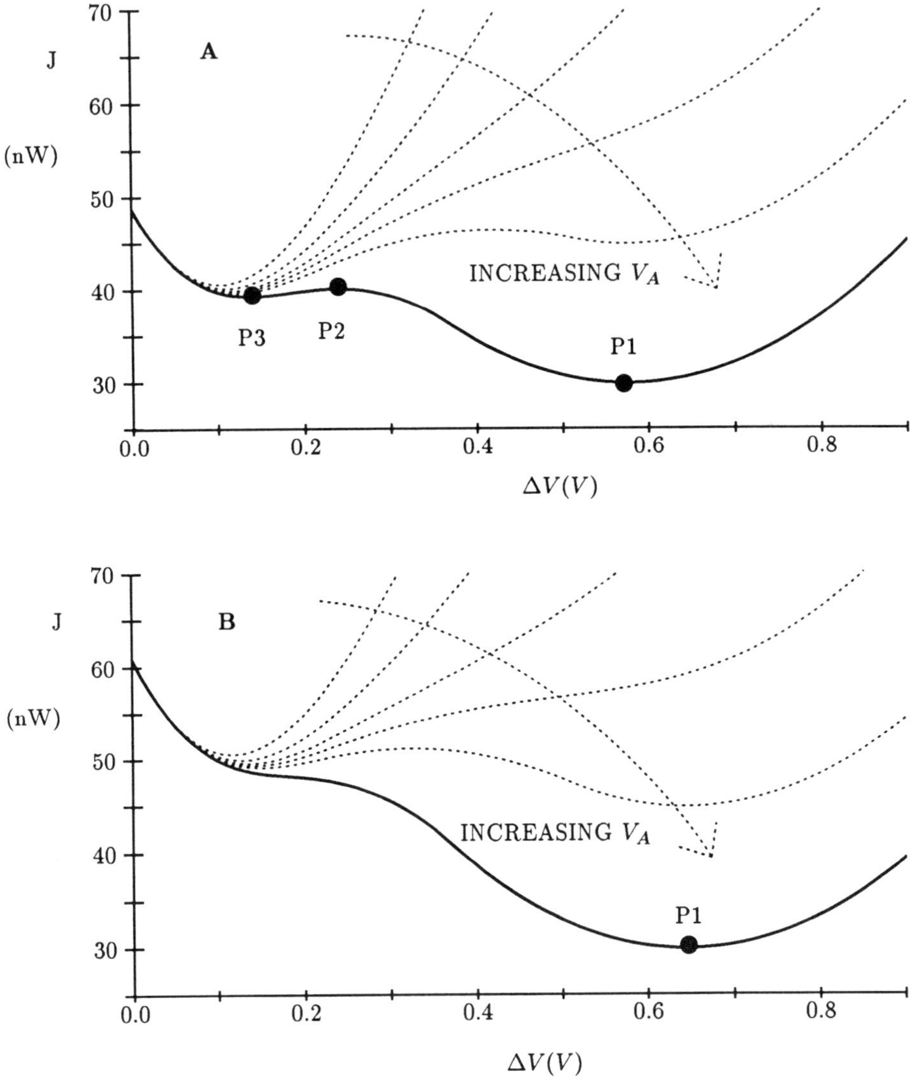

Figure 12 Computed total co-content from the I-V curves shown in Fig. 11. In plot A, P1 and P3 correspond to stable minima while P2 is an unstable maximum. In contrast, Plot B contains a single equilibrium point P1 that corresponds to a discontinuity. The dotted lines show the effect of increasing V_A.

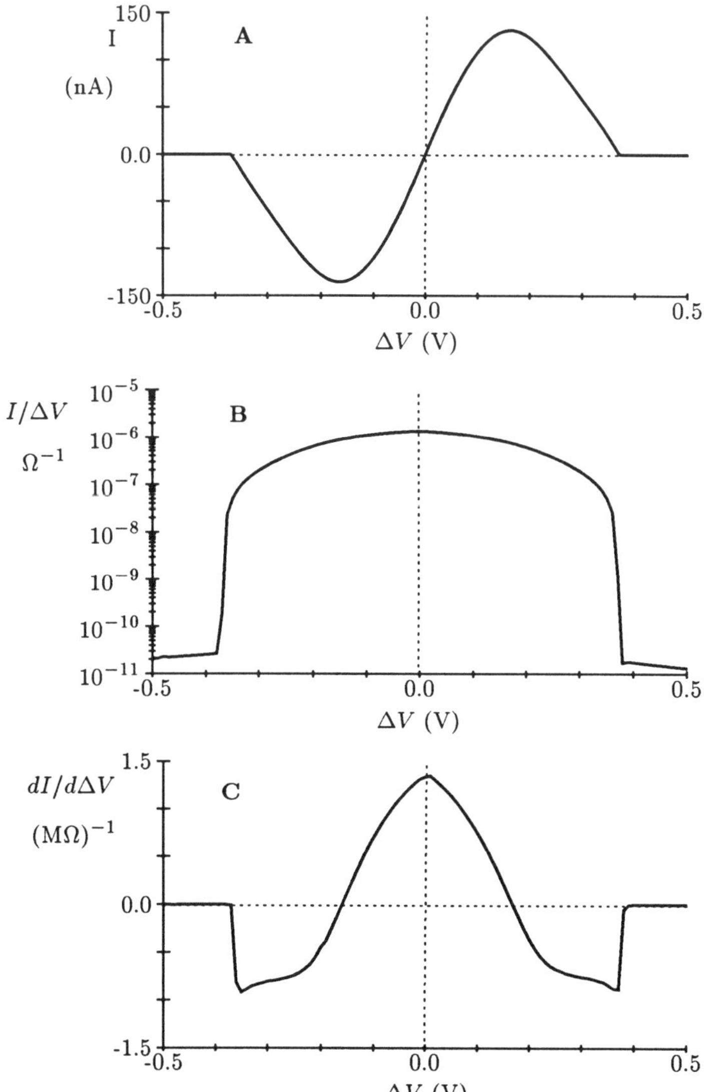

Figure 13 The I-V curve of the fuse measured in 10mV increments is shown in (A). (B) shows the numerically computed chord conductance, which is defined as $I/\Delta V$. Incremental conductance is defined to be $dI/d\Delta V$, which is the derivative of the I-V curve. (C) shows the incremental conductance computed using a two-point derivative approximation. Note the two regions of negative incremental conductance in (C).

will be *globally asymptotically stable*, i.e., for *any* voltage input and *any* initial condition it will converge to the unique stationary solution mentioned above. This conclusion holds even if positive, parasitic capacitances are distributed arbitrarily throughout the network, provided there are no inductors. This result assures us then that implementing the ideal, linear resistances dictated by standard regularization theory with Mead-type saturating resistances will not cause additional stationary solutions to appear.

2. Nonmonotonic Resistors

Now suppose the nonlinear resistors are externally *passive* (i.e., their I-V curves lie in the 1st and 3rd quadrants of the I-V plane) but are *incrementally active*, i.e., have regions of negative slope, as the resistive fuse in Fig. 3. Then there will in general exist a *number* of stationary network solutions for a given input image. If we further suppose that we can neglect the *internal* dynamics of the incrementally active resistor circuit, then for any voltage input and any initial condition the network will *not* oscillate indefinitely but must eventually settle to *some* stationary state. This conclusion also holds even if parasitic (positive) capacitances are distributed arbitrarily throughout the network, provided there are no inductors. This is a rather surprising result in view of the well-known instability problems with negative incremental resistance circuits.

3. Resistors with Internal Dynamics

The nonlinear resistors are of course multiple transistor circuits themselves and will inevitably have internal transient dynamics due to charge storage in transistors and parasitic wiring capacitance. Although each of the resistor circuits reported here is known to be stable in isolation, networks of such elements may, in principle, be unstable. This is an active research area, and many questions remain. Recent theoretical work (Wyatt and Standley, 1989; Standley and Wyatt, 1989; Standley, 1989) gives sufficient conditions for stability of such networks when the complex high-frequency dynamics are confined to the *linear* elements in any circuit consisting only of such linear elements, nonlinear memoryless resistors, and positive nonlinear capacitors. These results can be applied to yield *local* stability criteria for networks in which the resistor circuits are incrementally passive (such as Mead's saturating resistor) but have complex internal dynamics. But in their present form they are not applicable to networks in which the resistors are incrementally active (such as the resistive fuse) with internal dynamics.

The conclusions given in 1 and 2 above follow from well-established nonlinear network principles outlined below. Since the derivations follow with remarkable ease in these two cases, complete proofs are given.

We have sometimes found that experienced circuit designers can be deeply skeptical about the dynamic stability (non-oscillation) claim made above, and tunnel diode oscillator circuits are sometimes mentioned as counterexamples. It may be helpful to clarify what the precise result, Theorem II below, actually assumes. In the first place, it assumes an inductorless circuit, i.e., the only circuit elements allowed are positive (but possibly nonlinear) capacitors, ideal constant voltage sources, and nonlinear (possibly incrementally active) resistors. Thus oscillators that rely on inductors, even the distributed inductance in connecting wires, are not ruled out by the theorem. Note also that nonreciprocal building blocks, such as amplifiers, are not allowed under the assumptions, and that the individual resistors are assumed to have *no* internal dynamics of their own. Finally, the theorem does *not* assert that *every* stationary network solution is stable. Some will be unstable and some will be stable, but the network will eventually always settle to one of the latter.

The "no-inductors" assumption and the "no resistor dynamics" assumption are modelling approximations. Their appropriateness in a particular context is always open to question, and the issue can be settled for any given circuit only by experimentation. We note here that neglecting on-chip inductance has proven to be an excellent approximation in the analysis of many practical circuits, and that the nonlinear resistor circuits reported here are *intended* by the designer to operate as essentially memoryless resistors.

All the conclusions in 1 and 2 above follow easily from Tellegen's theorem, restated below for convenience (Tellegen, 1952; Penfield, Spence and Duinker, 1970; Chua, Desoer and Kuh, 1987).

4. Tellegen's Theorem

Assume we are given a network with sign conventions for branch voltages V_k and branch currents I_k such that the product $V_k \cdot I_k$ represents the power flowing *into* branch k. Then

$$\sum_{\text{all network branches}} V_k \cdot I_k = 0 \tag{18}$$

Furthermore, suppose x_k represents either V_k or any quantity derived from V_k such that at each instant the set of all x_k satisfies Kirchhoff's Voltage Law (KVL), i.e., the x_k sum to zero around any loop in the network. And suppose y_k represents either I_k or any quantity derived from I_k such that at each instant the set of all y_k's satisfies Kirchhoff's Current Law (KCL) i.e., the sum of the y_k's entering any node is zero (examples include $x_k = dV_k/dt, x_k(t) = V_k(t+3), y_k = \int I_k$, etc.). Then

$$\sum_{\text{all network branches}} x_k(t_1)y_k(t_2) = 0, \text{ for all } t_1, t_2. \tag{19}$$

Tellegen's theorem makes it very easy to show why the stationary solution to any network with incrementally passive resistors must be unique, as claimed in section 1.

5. Theorem I (Uniqueness)

There exists at most one solution for the resistor voltages and currents in any network of arbitrary topology consisting of strictly incrementally passive resistors and ideal voltage and current sources.

Proof: Suppose on the contrary there exist two such solutions, solution a and solution b (if more exist, pick any two). Let V_k^a and V_k^b denote the voltage across branch k in the two solutions, ΔV_k denote $V_k^b - V_k^a$, and let ΔI_k be defined similarly. Then the set of ΔV_k's satisfies KVL and the ΔI_k's satisfy KCL, so from eq. (19)

$$\sum_{\text{all resistors and sources}} \Delta V_k \cdot \Delta I_k = 0. \tag{20}$$

Since $V_k^a = V_k^b$ for all voltage sources and $I_k^a = I_k^b$ for all current sources, the product $\Delta V_k \cdot \Delta I_k$ vanishes for all source branches and eq. (20) reduces to

$$\sum_{\text{all resistors}} \Delta V_k \cdot \Delta I_k = 0. \tag{21}$$

But each resistor curve has positive slope by assumption, so $\Delta V_k \cdot \Delta I_k \geq 0$. Thus eq. (21) guarantees that $\Delta V_k = 0$ or $\Delta I_k = 0$ for each resistor. Therefore ΔV_k *and* ΔI_k *both* vanish since each resistor curve is assumed to be single-valued and invertible.
Q.E.D.

This theorem first appeared in Duffin (1947); see also Birkhoff and Diaz (1956). A more recent treatment can be found in Hasler (1986).

The non-oscillation claims in sections 1 and 2 follow with similar ease from Tellegen's theorem. The key quantity of interest is the *resistor co-content* of eq. (2) (see also Poggio and Koch, 1985). Thus, the reason nonlinear RC networks cannot exhibit unforced sustained oscillations, even if the resistors are incrementally active, is because $J_{total}(t)$ is always "running down," i.e. J_{total} acts (roughly speaking) as a Lyapunov function.

6. Theorem II (Stability)

Consider a network of arbitrary topology consisting of nonlinear voltage-controlled resistors, ideal time-invariant voltage sources, and nonlinear but positive capacitors described by $I_k = C_k(V_k)\frac{dV_k}{dt}$, with $C_k(V_k) > 0$ everywhere. Then J_{total} is strictly decreasing at each instant during any transient, i.e.,

$$\frac{dJ_{total}(t)}{dt} \leq 0, \tag{22}$$

and the inequality is strict except at equilibrium.

Proof: From Tellegen's theorem, eq. (19), we have

$$\sum_{\text{all network branches}} I_k(t)\frac{dV_k(t)}{dt} = 0. \tag{23}$$

For the voltage sources $\frac{dV_k(t)}{dt} = 0$, so these drop out of the sum in eq. (23), which now reads

$$\sum_{\text{all resistors}} I_k(t)\frac{dV_k(t)}{dt} + \sum_{\text{all capacitors}} I_k(t)\frac{dV_k(t)}{dt} = 0. \tag{24}$$

For each resistor,

$$I_k(t)\frac{dV_k(t)}{dt} = \frac{dJ_k(t)}{dt}, \tag{25}$$

which follows from eq. (2), using the chain rule for derivatives. Thus the first sum in eq. (24) is just $dJ_{total}(t)/dt$. And for each capacitor,

$$I_k(t)\frac{dV_k(t)}{dt} = C_k(V_k(t))\left(\frac{dV_k(t)}{dt}\right)^2 \geq 0. \tag{26}$$

The inequality (22) follows upon substituting eqs. (25) and (26) into (24). **Q.E.D.**

This theorem is a special case of results in (Brayton and Moser, 1964), but the proof given here is much more elementary.

If J_{total} is bounded from below and slopes upward for large values of the voltages, then Theorem II implies that the network will settle into a steady-state. A sufficient condition for this is that the I-V curve of all resistors in the network should lie somewhere in the interior of the 1st and 3rd quadrants for large values of ΔV.

Note that Theorem II rules out sustained oscillation because $J_{total}(t)$ would have to be periodic if the network state were periodic, and this is impossible since $dJ_{total}/dt \leq 0$, with equality only at equilibrium. However, J_{total} does not necessarily meet all the standard criteria for a Lyapunov function since its *shape* is essentially arbitrary. It is easy to show that J is convex if and only if the resistors are all incrementally passive. With incrementally active resistors such as resistive fuses, J can have many *local minima*, which are then the (locally) *stable* equilibria of the network. In the case of positive linear resistors, Theorem 2 has the special interpretation that the total *dissipated power* decreases monotonically during transients in any RC circuit with voltage sources, even if the capacitors are nonlinear. In this linear case the co-content (and the total power) are convex functions of those voltages that are not constrained by the sources, so the local minimum to which the network converges is in fact the global minimum of the dissipated power, subject to the source constraint. Stripped of all dynamics, the static version of this statement is known as *Maxwell's Minimum Heat Theorem* (Maxwell, 1891).

CONCLUSION

We have successfully demonstrated in this manuscript for the first time a simple and elegant analog circuit implementation of the line discontinuities of Geman and Geman (1984) and of the graduated non-convexity algorithm of Blake and Zisserman (1987). We only report on the experimental data for an 8 pixel 1-D circuit. We have sent out a 20 by 20 pixel 2-D version of this network to MOSIS for fabrication. We previously demonstrated a 48 by 48 pixel circuit implementing smooth surface interpolation (Luo, Koch and Mead, 1988). This work can be extended to include 2nd order or thin-plate surface interpolation (Harris, 1989), where the energy functional embodies the discretized square of the ∇^2 operator. Computer simulations have shown that detection of discontinuities in surface orientation, such as occurring along creases, is feasible in problems such as edge detection and surface interpolation (Blake and Zisserman, 1987; Liu and Harris, 1989) and can be incorporated into our thin-plate interpolation circuits (Harris, 1989).

We thus have all the elementary circuit elements in hand—phototransistors for on-chip image acquisition (Mead, 1989), resistive networks for smoothing, and resistive fuses for detecting discontinuities—to design analog, resistive network chips to compute the 2-D optical flow field in the presence of motion discontinuities, the depth and depth discontinuities in 2-D images as well as intensity discontinuities.

Acknowledgements

We thank Carver Mead for laying the framework upon which we have built our research. Our theoretical ideas would never have been ported into silicon without him. Hewlett-Packard provided computing support. All chips were fabricated through MOSIS with DARPA's support. Research in the laboratory of C.K. is supported by a National Science Foundation grant IST-8700064, Office of Naval Research Young Investigator and National Science Foundation Presidential Young Investigator Awards, the James S. McDonnell Foundation, Rockwell International Science Center and the Hughes Aircraft Artificial Intelligence Center. J.W. acknowledges DARPA Contract No. N00014-87-K-0825, National Science Foundation Contract No. MIP- 8814612, and the DuPont Corporation. We thank David Standley for his careful reading of this document and Bimal Mathur for inspiring the high-gain fuse circuit.

References

Birkhoff, G. and Diaz, J. B. (1956). Nonlinear network problems. *Quart. Appl. Math.* **13**:431–443.

Blake, A. (1989). Comparison of the efficiency of deterministic and stochastic algorithms for visual reconstruction. *IEEE Trans. Pattern Anal. Mach. Intell.* **11**:2–12.

Blake, A. and Zisserman, A. (1987). *Visual Reconstruction.* Cambridge, MA: MIT Press.

Brayton, R. K. and Moser, J. K. (1964). A theory of nonlinear networks—I, II. *Quart. Appl. Math.* **22(1)**:1–33 (April) and **22(2)**:81–104 (July).

Chua, L. O., Desoer, C. A., and Kuh, E. S. (1987). *Linear and Nonlinear Circuits.* New York: McGraw-Hill, pp. 23–34.

Duffin, R. J. (1947). Nonlinear networks IIa. *Bull. Amer. Math. Soc.* **53**:963–971.

Geman, S. and Geman, D. (1984). Stochastic relaxation, Gibbs distribution and the Bayesian restoration of images. *IEEE Trans. Pattern Anal. Mach. Intell.* **6**:721–741.

Grimson, W. E. L. (1981). *From Images to Surfaces.* Cambridge, MA: MIT Press.

Harris, J. G. (1989). An analog VLSI chip for thin plate surface interpolation. In *Neural Information Processing Systems*, ed. D. Touretzky. Palo Alto: Morgan Kaufmann.

Harris, J. G. and Koch, C. (1989). Resistive fuses: circuit implementations of line discontinuities in vision. *Snowbird Neural Network Workshop*, April 4–7.

Harris, J. G., Koch, C., Staats, E., Luo, J. and Wyatt, J. (1989). Analog hardware for detecting discontinuities in early vision: computational justification and VLSI circuits, in preparation.

Hasler, M. and Neirynck, J., (1986). *Nonlinear Circuits.* Norwood, MA: Artech House Inc., pp. 172–173.

Hildreth, E. C. (1984). *The Measurement of Visual Motion.* Cambridge, MA: MIT Press.

Hopfield, J. J. and Tank, D. W. (1985). Neural computation in optimization problems. *Biol. Cybern.* **52**:141–152.

Horn, B. K. P. (1989). Parallel networks for machine vision. *Artif. Intell. Lab. Memo No.* **1071** (MIT, Cambridge).

Horn, B. K. P. and Schunck, B. G. (1981). Determining optical flow. *Artif. Intell.* **17**:185–203.

Ikeuchi, K. and Horn, B. K. P. (1981). Numerical shape from shading and occluding boundaries. *Artif. Intell.* **17**:141–184.

Hutchinson, J., Koch, C., Luo, J., and Mead, C. (1988). Computing motion using analog and binary resistive networks. *IEEE Computer* **21**:52–63.

Ikeuchi, K. and Horn, B. K. P. (1981). Numerical shape from shading and occluding boundaries. *Artif. Intell.* **17**:141–184.

Koch, C., Marroquin, J., and Yuille, A. (1986). Analog "neuronal" networks in early vision. *Proc. Natl. Acad. Sci. USA* **83**:4263–4267.

Koch, C. (1989). Seeing chips: analog VLSI circuits for computer vision. *Neural Computation* **1**:184–200.

Liu, S. C. and Harris, J. G. (1989). Generalized smoothing networks in solving early vision problems. *Computer Vision and Pattern Recognition Conference.*

Luo, J., Koch, C., and Mead, C. (1988). An experimental subthreshold, analog CMOS two-dimensional surface interpolation circuit. *Neural Information Processing Systems Conference*, Denver, November.

Marr, D. and Poggio, T. (1976). Cooperative computation of stereo disparity. *Science* **194**:283–287.

Marroquin, J., Mitter, S., and Poggio, T. (1987). Probabilistic solution of ill-posed problems in computational vision. *J. Am. Statistic Assoc.* **82**:76–89.

Maxwell, J. C. (1891). *A Treatise on Electricity and Magnetism*, 3rd ed., Vol. I, pp. 407–408. Republished by New York: Dover Publications, 1954.

Mead, C. A. (1985). A sensitive electronic photoreceptor. In *1985 Chapel Hill Conference on Very Large Scale Integration*, pp. 463-471.

Mead, C. A. (1989). *Analog VLSI and Neural Systems.* Reading: Addison-Wesley.

Millar, W. (1951). Some general theorems for non-linear systems possessing resistance. *Phil. Mag.* **42**:1150–1160.

Nagel, H. H. (1987). On the estimation of optical flow: relations between different approaches and some new results. *Artif. Intell.* **33**:299–324.

Penfield, P., Jr., Spence, R., and Duinker, S. (1970). *Tellegen's Theorem and Electrical Networks*, Cambridge, MA: MIT Press.

Perona, P. and Malik, J. (1988). A network for multiscale image segmentation. *Proc. 1988 IEEE Int. Symp. on Circuits and Systems*, Espoo, Finland, June, pp. 2565–2568.

Poggio, T., Gamble, E. B., and Little, J. J. (1988). Parallel integration of vision modules. *Science* **242**:436–440.

Poggio, T. and Koch, C. (1985). Ill-posed problems in early vision: from computational theory to analogue networks. *Proc. R. Soc. Lond. B* **226**:303–323.

Poggio, T., Torre, V., and Koch, C. (1985). Computational vision and regularization theory. *Nature* **317**:314–319.

Poggio, T., Voorhees, H., and Yuille, A. (1986). A regularized solution to edge detection. *Artif. Intell. Lab Memo* **No. 833** (MIT, Cambridge).

Sivilotti, M.A., Mahowald, M.A. and Mead, C.A., Real-time visual computation using analog CMOS processing arrays. In: *1987 Stanford Conf. VLSI*, pp. 295-312 (MIT Press, Cambridge, 1987).

Standley, D. L., and Wyatt, J. L., Jr. (1989). Stability criterion for lateral inhibition and related networks that is robust in the presence of integrated circuit parasitics. In *IEEE Trans. Circuits and Systems* **36**, May., pp. 675–681

Standley, D. L. (1989). Design criteria extensions for stable lateral inhibition networks in the presence of circuit parasitics. *Proc. 1989 IEEE Int. Symp. on Circuits and Systems*, Portland, Oregon, May, pp. 837–840.

Tellegen, B. D. H. (1952). A general network theorem, with applications. *Phillips Research Reports* **7**:259–269.

Terzopoulos, D. (1983). Multilevel computational processes for visual surface reconstruction. *Comp. Vision Graph. Image Proc.* **24:** 52–96.

Terzopoulos, D. (1986). Regularization of inverse problems involving discontinuities. *IEEE Trans. Pattern Anal. Machine Intell.* **8**:413–424 (1986).

Wyatt, J. L., Jr. and Standley, D. L. (1989). Criteria for robust stability in a class of lateral inhibition networks coupled through resistive grids. *Neural Computation* **1**:58–67.

3

CMOS INTEGRATION OF HERAULT-JUTTEN CELLS FOR SEPARATION OF SOURCES

Eric A.Vittoz — CSEM, Neuchâtel

Xavier Arreguit — EPF, Lausanne

INTRODUCTION

Let us consider an array of n unknown independent sources $X_i(t)$ which, at any time, are only observable indirectly through the n signals $E_i(t)$ obtained by an unknown linear combination of the X_i (Fig.1). In vectorial notation:

$$\underline{E}(t) = [A]\,\underline{X}(t) \tag{1}$$

where [A] is a square n-matrix

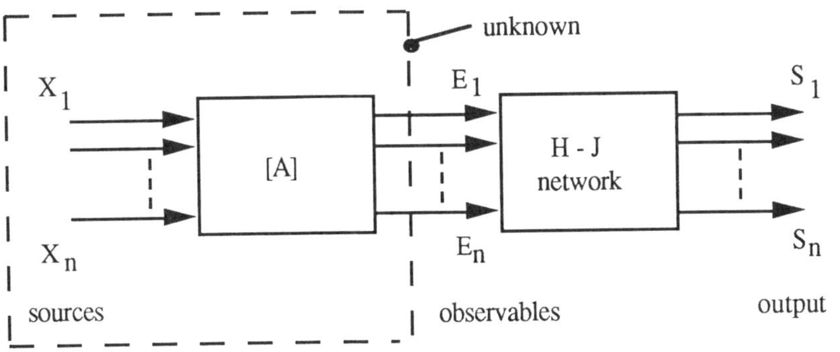

Fig.1 Separation of sources

The problem is to restore the primary signals $\underline{X}$ without a priori knowledge of the mixing matrix [A]. This problem was first addressed by J.Hérault and C.Jutten [1] after they found some evidence that the informations about speed and position of body joints are mixed up before being sent to the brain by two different types of nerves. The brain is however perfectly capable of separating speed and position.

A comparable engineering situation is the case of n sensors, each with an unknown (and possibly slowly variable) sensitivity to n independent variables. The problem is then to extract these variables from the signals provided by the sensors.

Hérault and Jutten have proposed a network [1],[2] which carries out this task, by transforming the measured vector $\underline{E}$ into a new vector $\underline{S}$ with independent components S_i. Each S_i is therefore proportional to only one of the sources X_i.

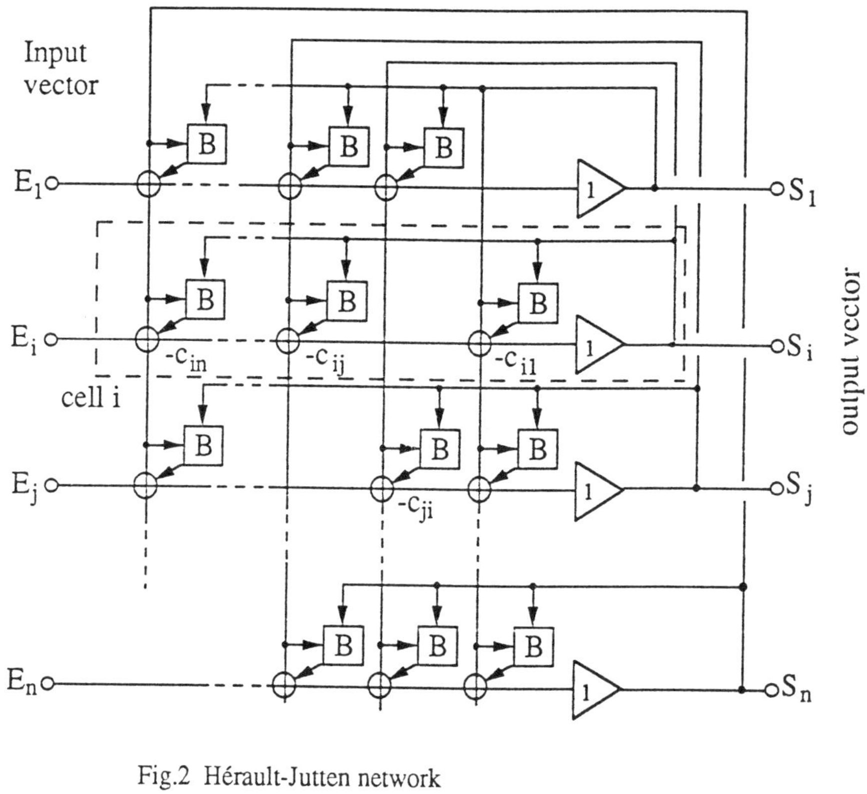

Fig.2 Hérault-Jutten network

The block diagram of this network is shown in Fig.2. It is a (almost) fully interconnected network of n cells (or neurons) providing n output values S_i. Each cell i is driven by the sum of one input E_i and output S_j of all the other cells, each weighted by a negative factor $-c_{ij}$ (synaptic weight).

The value of c_{ij} is adaptable by means of a local law (which only depends on the values of S_i and S_j) implemented by the nonlinear adaptation block B. The whole network is otherwise linear, with

$$S_i = E_i - \sum_{j \neq i} c_{ij} S_j \tag{2}$$

This structure and its behaviour are analyzed in details in the Ph.D.Thesis of C.Jutten [3], which has provided most of the theoretical material used here. This paper will describe an experimental CMOS analog implementation of one cell.

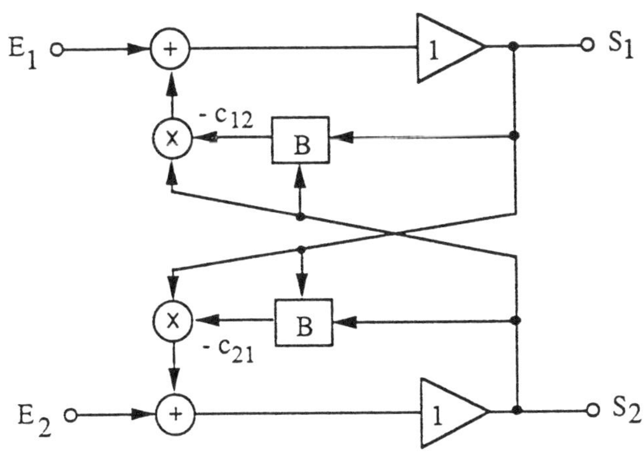

Fig.3 Detailed structure of a 2-cell H-J network

SUMMARY OF THE BASIC THEORY

To simplify the explanations, let us consider the 2-cell network represented in Fig.3. In this case:

$$S_1 = E_1 - c_{12} S_2 \tag{3}$$

$$S_2 = E_2 - c_{21} S_1 \tag{4}$$

The solution of which is:

$$S_1 = \frac{E_1 - c_{12} E_2}{1 - c_{12} c_{21}} \quad (5) \qquad \text{and} \qquad S_2 = \frac{E_2 - c_{21} E_1}{1 - c_{21} c_{12}} \quad (6)$$

Let us now express the fact that E_1 and E_2 are linear combinations of the original sources X_1 and X_2:

$$E_1 = a_{11} X_1 + a_{12} X_2 \tag{7}$$

$$E_2 = a_{21}.X_1 + a_{22} X_2 \tag{8}$$

where the a_{ij} are the coefficients of the mixing matrix [A]. The output signals are then given by:

$$S_1 = \frac{(a_{11} - c_{12}\, a_{21})\, X_1 + (a_{12} - c_{12}\, a_{22})\, X_2}{1 - c_{12}\, c_{21}} \tag{9}$$

$$S_2 = \frac{(a_{21} - c_{21}\, a_{11})\, X_1 + (a_{22} - c_{21}\, a_{12})\, X_2}{1 - c_{12}\, c_{21}} \tag{10}$$

Each output may be made to depend on only one source by cancelling the contribution of the other. Two solutions are in principle possible:

$c_{12} = \frac{a_{12}}{a_{22}}$ and $c_{21} = \frac{a_{21}}{a_{11}}$ which provides:

$$S_1 = a_{11} X_1 \quad \text{and} \quad S_2 = a_{22} X_2 \tag{11}$$

$c_{12} = \frac{a_{11}}{a_{21}}$ and $c_{21} = \frac{a_{22}}{a_{12}}$ which provides:

$$S_1 = a_{12} X_2 \quad \text{and} \quad S_2 = a_{21} X_1 \tag{12}$$

However, the system is a closed loop and a stable solution is only possible if $c_{12}.c_{21}<1$. As a result, the extraction of source X_i must be done by the channel which has the larger relative content of this source.

The problem is now to find a way to adapt the coefficients c_{ij} to the value required for separation without knowing the coefficients a_{ij} of the mixing matrix [A]. The only clue available is that the unknown sources X_i are independent. Therefore, after separation, the output signals S_i must be independent as well.

For the sake of simplicity, let us assume that the average values of all sources X_i are zero:

$$E(X_i) = 0 \text{ and therefore } E(S_i) = 0 \tag{14}$$

where E(x) is the mathematical expectation of x. The condition for the output signals to be decorrelated (zero covariance) is expressed as

$$E(S_1.S_2) = 0 \tag{15}$$

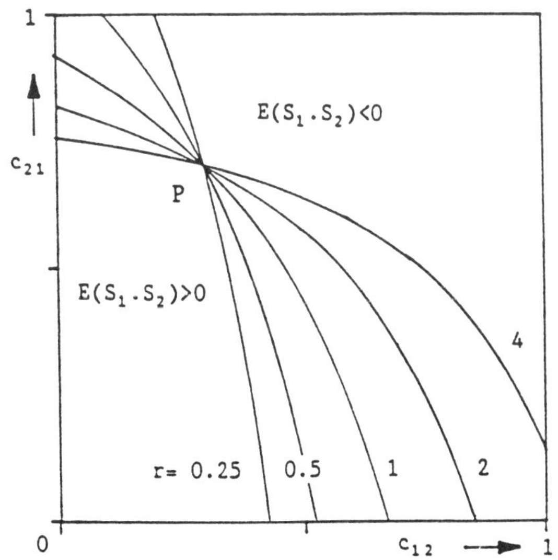

Fig.4 Solution P and loci of zero covariance

Introducing expressions (9) and (10) of the output signals and using the fact that the original sources are not correlated provides a relation between c_{12} and c_{21} [3]. As shown in Fig.4, this locus of zero covariance depends on the ratio r of the variances of the sources

$$r = \frac{E(X_1{}^2)}{E(X_2{}^2)} \tag{16}$$

except at point P which corresponds to the solution of the problem (relation (11) or (12)). Furthermore, the covariance is positive below the curve and negative above. Therefore, the synaptic weights will converge to values ensuring decorrelation if

$$\frac{d c_{12}}{d t} = \frac{d c_{21}}{d t} = a \; E(S_1 \, S_2) \tag{17}$$

By choosing the gain a sufficiently small, the weights only change very slowly and the expectation may be replaced by the intantaneous value.

Such an adaptation law is however not sufficient to reach the required solution P because it only achieves decorrelation of the output signals. What is needed to reach P is independence, which is a much stronger requirement.

It can be shown that the independence of S_1 and S_2 implies

$$E\left(S_1{}^k\, S_2{}^l\right) = 0 \quad \text{for any odd value of k and l} \tag{18}$$

On this basis, Hérault and Jutten have proposed an adaptation law of the form

$$\frac{d\,c_{12}}{d\,t} = a\ f(S_1)\ g(S_2) \tag{19}$$

$$\frac{d\,c_{21}}{d\,t} = a\ f(S_2)\ g(S_1) \tag{20}$$

where f and g are odd functions which contain the necessary odd power terms to force condition (18). They must be different to permit the asymmetrical variation of c_{12} and c_{21} necessary to reach P from any initial condition. For a practical analog implementation the choice of these functions will be dictated by their realizability (opportunistic approach defined by C.Mead (4)). However, interesting clues may be obtained by a priori simulations with simple functions.

NUMERICAL SIMULATIONS

These are easily carried out by calculating S_1 and S_2 from equation (5) and (6) for the present value of input signals E_1 and E_2. These values are then introduced in (19) and (20) to calculate the increment of c_{12} and c_{21}, which can be plotted before going to the next cycle with the next values of E_1 and E_2. In the following simulations, the sources X_1 and X_2 are sequences of independent random values with probability constant between X_{max} and $-X_{max}$ and zero outside. The mixing matrix is kept constant with:

$$a_{11}=1,\ a_{12}=0.3,\ a_{21}=0.7 \text{ and } a_{22}=1$$

From (11) the solution P is then $c_{12}=0.3$, $c_{21}=0.7$

Figures 5 to 7 show the trajectories in the (c_{12},c_{21}) plane for 3 different pairs of simple functions f and g. The convergence is facilitated by choosing f and g both nonlinear with opposite curvature signs. Starting at the origin, the trajectory approaches quickly the locus of zero covariance before turning more slowly towards point P. When it is started somewhere on this locus, the trajectory follows it until solution P is reached.

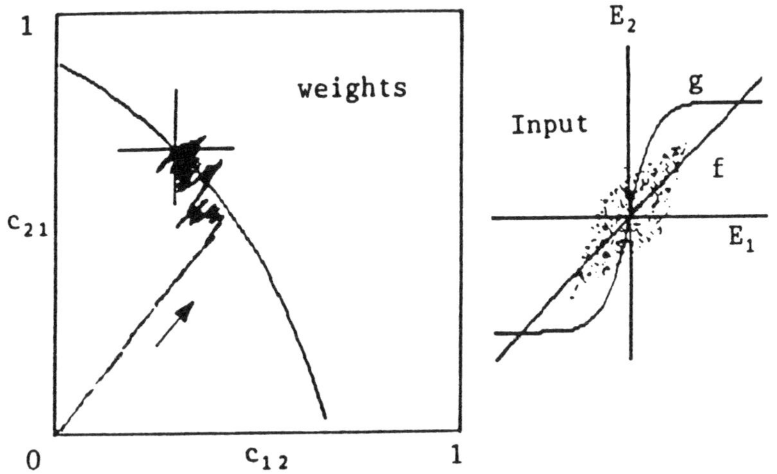

Fig.5 Convergence with f(x)=x and g(x)=tanh(x)

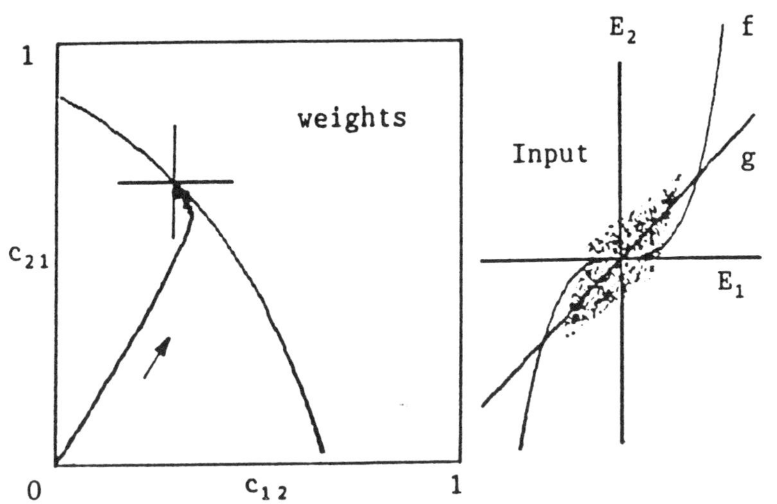

Fig.6 Convergence with $f(x)=x^3$ and g(x)=x

Fig.8 and 9 show the results for nonlinear functions with the same curvature sign. The trajectory reaches the locus of zero covariance but then moves away from P.

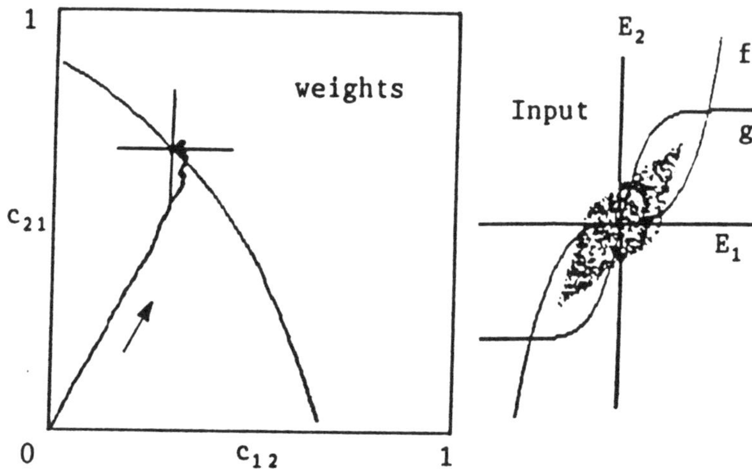

Fig.7 Convergence with $f(x)=x^3$ and $g(x)=\tanh(x)$

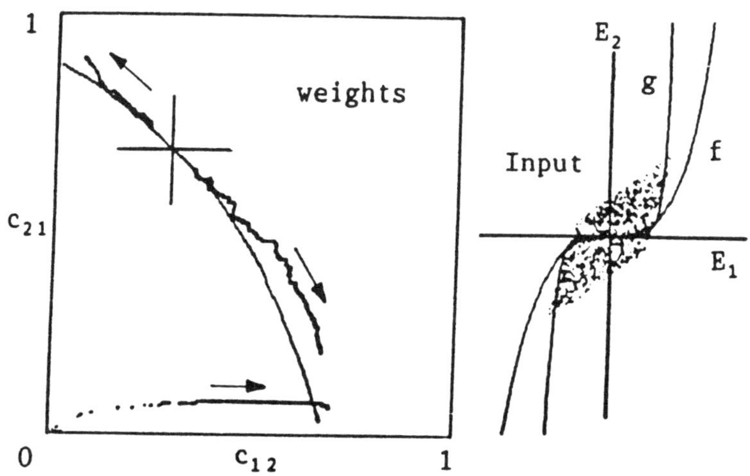

Fig.8 Divergence with $f(x)=x^3$ and $g(x)=x^7$

Figure 10 shows the results obtained when X_1 and X_2 are binary distributions with 2 equiprobable states X_{max} and $-X_{max}$. The simulation has been started on the locus of zero covariance but far from solution P. The evolution at the output is also shown in the (S_1, S_2) plane.

A particular problem arises when the statistical distribution of amplitudes of the sources is gaussian. Indeed, decorrelation is then equivalent to independence and the system does not have any clue to find solution P on the locus of zero covariance. This is clearly visible in the simulation of Fig.11. However, in practice, the sources are never exactly gaussian in a given period of time. The system can still converge as will be shown in the practical implementation.

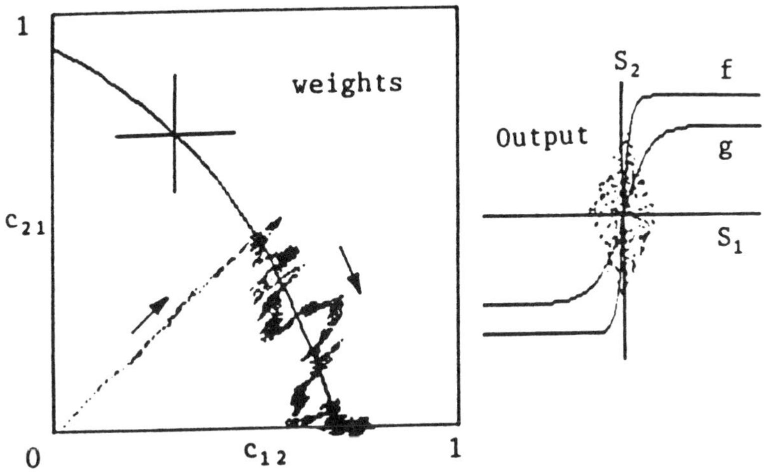

Fig 9 Divergence with f(x)=tanh(3x) and g(x)=tanh(x).

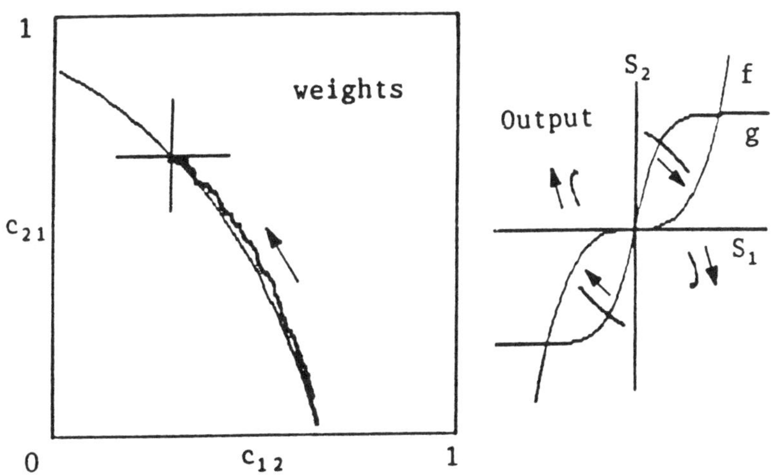

Fig.10 $f(x)=x^3$ and g(x)=tanh(x); convergence with binary sources with 2 equiprobable states. The locus in output plane S_1-S_2 is also shown.

The principle involved may be explained qualitatively by saying that the system forces a symmetrical distribution of amplitudes in the output plane (S_1, S_2). Indeed, because f and g are both odd functions and the variables have zero mean values, c_{12}

and c_{21} (given by (19) and (20)) stop varying when biaxial symmetry is achieved in the (S_1, S_2) plane.

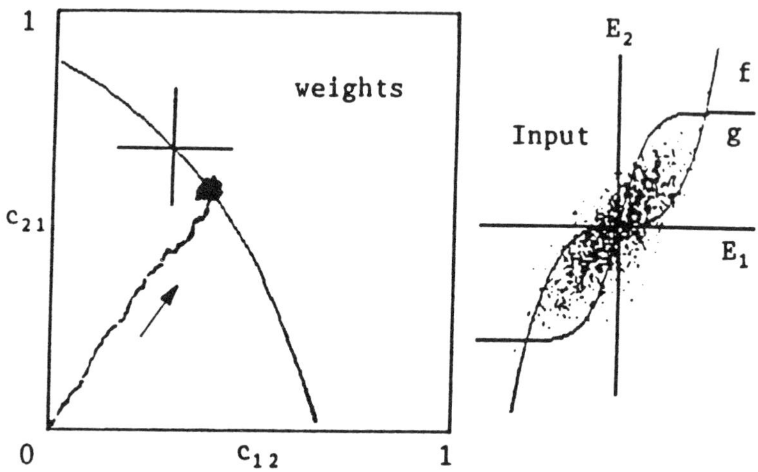

Fig.11 Simulation with $f(x)=x^3$ and $g(x)=\tanh(x)$. Apparent lack of convergence with gaussian sources.The system still converges in practical implementations.

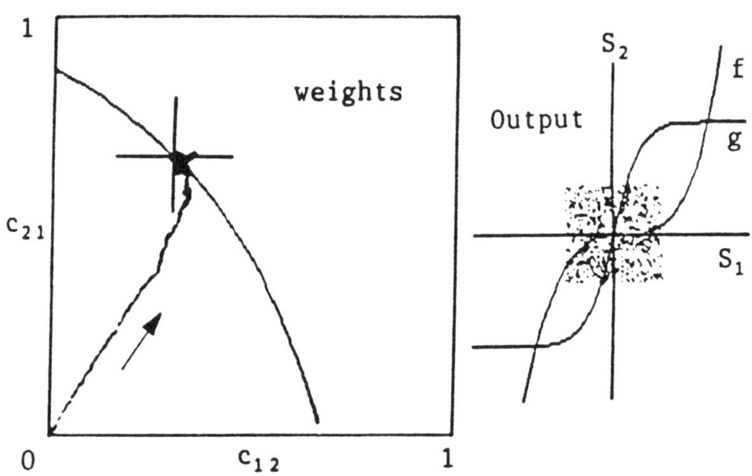

Fig.12 The system still converges perfectly with an offset of 40% of the RMS value of the sources in one of the functions.

In Fig.12, the simulation shows that the system is not affected by a large asymmetry in one of the functions. This is because a single axial symmetry in plane (S_1, S_2) is sufficient to achieve equilibrium. As a matter of fact, if one of the functions is perfectly symmetrical, the other may contain any even component without affecting the system.

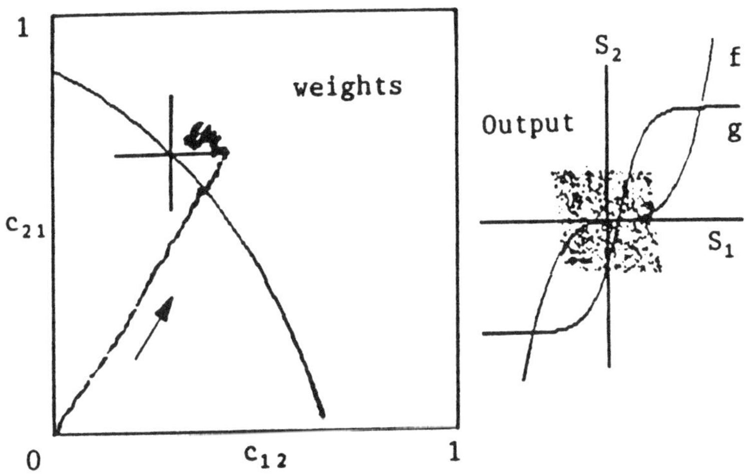

Fig.13 The convergence is no more acceptable when the two functions are offset by 40% of the RMS value of the sources.

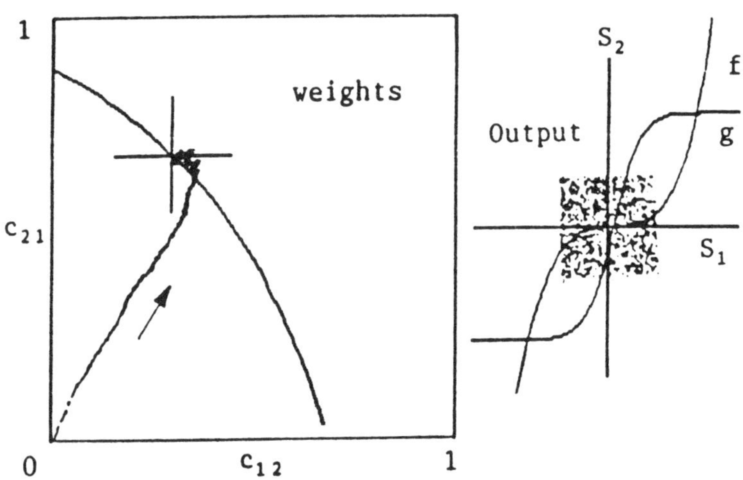

Fig.14 Acceptable solution when both functions are offset by 20% of the RMS value of the sources.

Fig.13 shows the result with large asymmetries in both functions (each corresponding to 40% of the RMS values of S_1 and S_2). The system does not converge to P anymore and separation is not carried out properly. Separation becomes perfectly acceptable as soon as these asymmetries are reduced to no more than 20%, as demonstrated in Fig.14. The system is thus not very sensitive to asymmetries. This is a very important feature for analog implementations in which mismatches of components must be considered. The algorithm is very robust. It works properly with a wide range of nonlinear functions, provided they are of opposite curvature sign.

ANALOG CMOS IMPLEMENTATION

The block diagram of the realized cell is shown in Fig.15. Unity gain amplification is obtained by an operational amplifier OA and 2 identical transistors T_{i0}, T_{iR} with same gate voltage V_R. This voltage must be much larger than the peak value of input E_i to ensure linear operation.

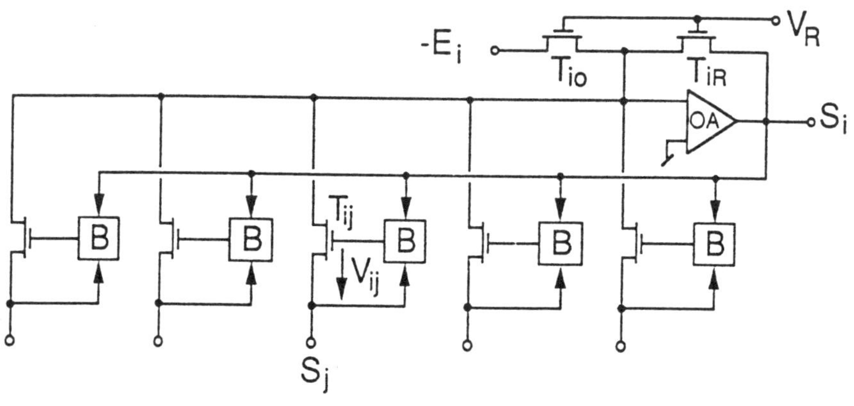

Fig.15 Block diagram of one cell.

The output signals S_j of all the other cells are added and weighted by transistors T_{ij} with gate voltage V_{ij}. The corresponding synaptic weights for all transistors identical are given by

$$c_{ij} = \frac{V_{ij} - V_T}{V_R - V_T} \tag{21}$$

where V_T is the gate threshold voltage of the transistors.

The operational amplifier (Fig.16) is just an elementary transconductance amplifier followed by a common drain stage. Bias currents I_{p1} and I_{p2} are separately adjustable to provide more flexibility in this experimental circuit.

Each voltage V_{ij} is generated by a block B in such a way as to make the local output signal S_i statistically independent of signals S_j from other cells, according to the principle described above. The circuit realization of B must be as simple as possible while carrying out the task. It should exploit the intrinsic nonlinear characteristics of the transistors to create the required functions f and g.

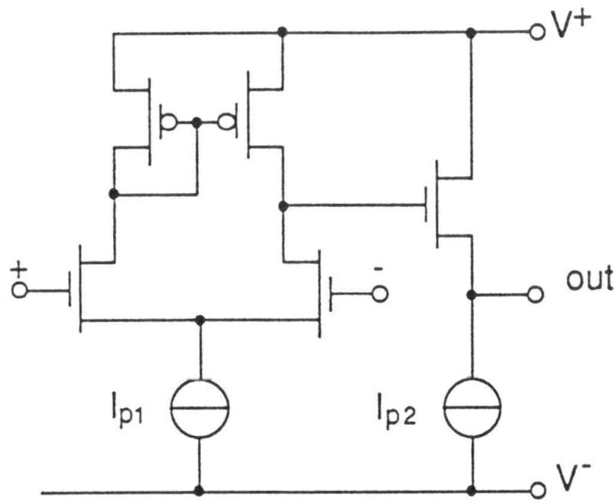

Fig.16 Operational amplifier.

Model of the transistor.

The drain current I_D of a transistor in saturation can be analytically modelled by the following equations [5]:

weak inversion (small currents):

$$I_D = I_s \exp \frac{V_G}{nU_T} \qquad \text{for } I_D \ll I_s \tag{22}$$

strong inversion (large currents):

$$I_D = I_s \left[\frac{V_G - V_T}{2nU_T}\right]^2 \qquad \text{for } I_D \gg I_s \tag{23}$$

where:

$$I_s = 2n\, \mu Cox\, U_T^2 \frac{W}{L} \tag{24}$$

$$U_T = \frac{kT}{q} \tag{25}$$

V_T is the threshold value of gate voltage V_G and n is the body effect which also affects the slope in weak inversion. C_{ox} is the gate oxide capacitance per unit area and μ is the charge carrier mobility. W/L is the width to length ratio of the transistor channel.

Moderate inversion:

when I_D is the same order of magnitude as I_s, operation is in moderate inversion [6]. If necessary, the drain current may then be obtained by interpolation between weak and strong inversion [7].

Calculations can be simplified by normalizing

all voltages to nU_T: $\quad v = V / nU_T \qquad s_i = S_i / nU_T$ (26)
all currents to I_s : $\quad i = I / I_s$ (27)

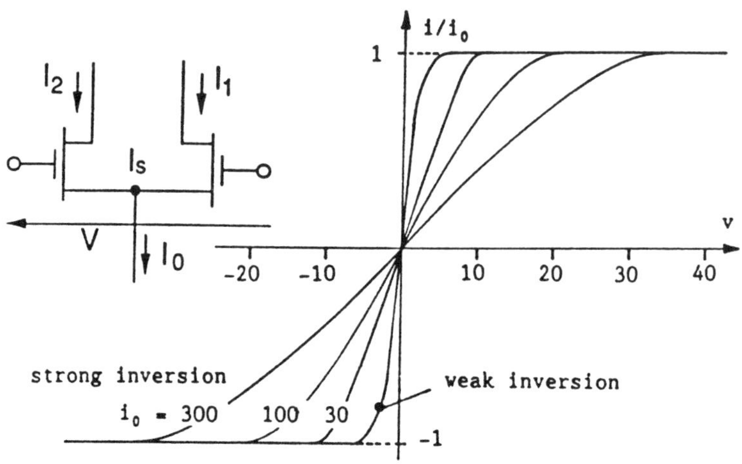

Fig.17 Differential pair and its transfer characteristics.

Differential pair.

The basic circuit brick of our implementation is the differential pair illustrated in Fig.17. Using the model above, the transfer characteristics of this circuit may be expressed as:

in weak inversion

$$i_{1(2)} = \frac{i_0}{1 + \exp\left(-(+)\, v\right)} \tag{28}$$

$$i = i_1 - i_2 = i_0 \tanh(v/2) \tag{29}$$

in strong inversion

$$i_{1(2)} = \frac{i_0}{2}\left[1 +(-)\sqrt{\left(8 - \frac{v^2}{i_0}\right)/\, i_0}\;\frac{v}{4}\right] \qquad (30)$$

for $v^2 \leq 4\, i_0$

$$i = i_1 - i_2 = \sqrt{8 i_0 - v^2}\;\frac{v}{4} \qquad (31)$$

$$i = i_1 = +i_0 \qquad \text{for } v \geq 2\sqrt{i_0} \qquad (32)$$

$$i = -i_2 = -i_0 \qquad \text{for } v \leq -2\sqrt{i_0} \qquad (33)$$

Relations (29) and (31) to (33) are represented in Fig. 17. As shown by equation (29), weak inversion can be used to implement a saturating function g which is multiplied by the tail current i_0. This current will have to be produced by the second function f of opposite curvature sign to implement the Hérault-Jutten algorithm. The situation is a little more complicated in strong inversion since i_0 does not have a purely multiplying effect.

Function f.

A good way to realize a function which has an increasing slope instead of saturating is to use a circuit with positive feedback. Such a circuit is shown in Fig.18 [8]. It is based on a differential pair T_f. The tail current I_0 of this pair is increased by A times its output current I_1:

$$I_0 = I_B + A.I_1 \qquad (34)$$

where I_B is a small fixed bias current.

The transfer function of the circuit can be calculated by introducing this relation into those of the differential pair, with all voltages and currents normalized according to (26) and (27). The results are:

in weak inversion

$$i_1 = \frac{i_B}{1 - A + \exp(-v)} \qquad (35)$$

Hence, the current obtained from a push-pull configuration will be:

$$i = i_1(v) - i_1(-v) = \frac{(\exp(v) - \exp(-v))\, i_B}{\left(1 - A + \exp(v)\right)\left(1 - A + \exp(-v)\right)} \qquad (36)$$

or, for feedback factor A=1:

$$i = 2\, i_B \sinh(v) \qquad (37)$$

in strong inversion (assuming $i_1 >> i_B$)

$$\begin{aligned} i_1 &= K v^2 && \text{for } v > 0 \\ i_1 &= 0 && \text{for } v < 0 \end{aligned} \qquad (38)$$

push-pull:

$$i = \mathrm{sgn}(v)\, K v^2 \qquad (39)$$

where

$$K = \frac{1}{4\left(A - \sqrt{A^2 - (2-A)^2}\right)} \qquad (40)$$

Thus, both modes of operation realize functions with the required overall shape (curvature positive for v positive and negative for v negative).

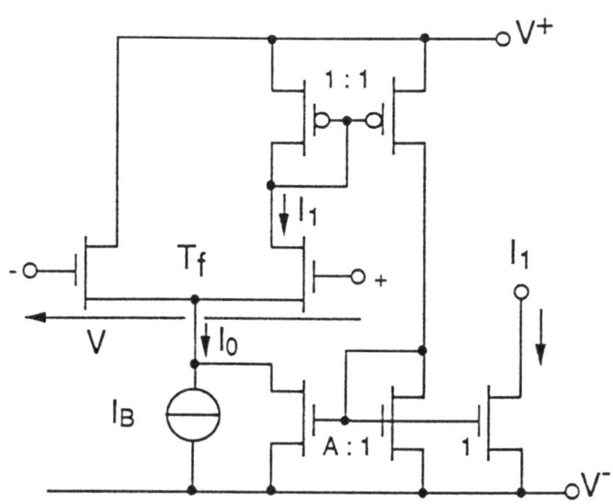

Fig.18 Principle used for generating function f.

Adaptation block B.

Each complete block B can now be built by combining 2 circuits of this type to realize the function f handling the two polarities of S_i, each followed by a differential pair to implement the multiplication by function g. As shown in the circuit diagram of Fig.19, these 2 contributions of current are then added and integrated in a capacitor C to create the gate control voltage V_{ij} of synaptic weight c_{ij}. Equilibrium is reached when no current is flowing in or out of C.

The adaptation law implemented by this circuit may be calculated by combining the results previously carried out for the various subcircuits. It depends on the mode of operation of differential pair T_f (specific current I_{sf}) in the f-function generator and on that of pair T_g (specific current I_{sg}) in the g-function generator.

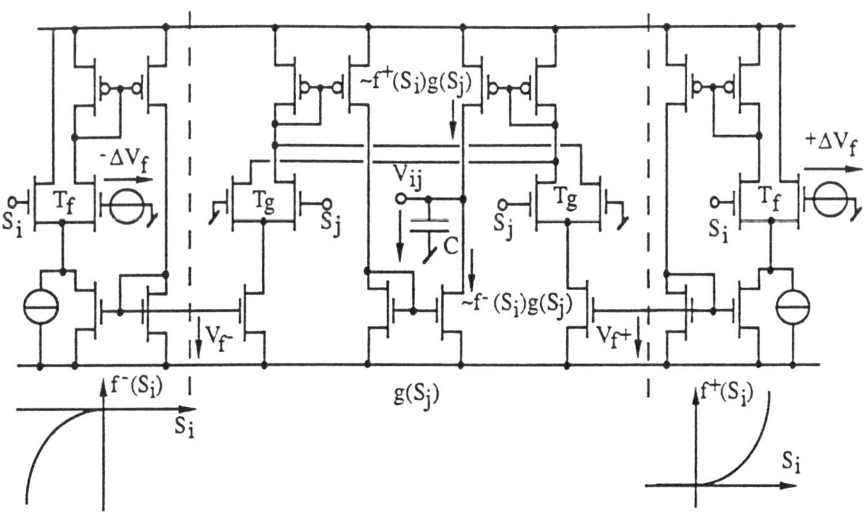

Fig.19 Circuit diagram of adaptation block B.

f and g in weak inversion

The maximum level of all the currents in block B is controlled by the f-function generator. It depends therefore on the maximum value of the local output S_i. According to relations (22) and (37), the differential pair T_f in this generator will stay in weak inversion as long as

$$\sinh(s_i) \ll I_{sf}/2I_B \tag{41}$$

The differential pair T_g in the g-function generator will stay in weak inversion as well if

$$R = M\, I_{sf}/I_{sg} \tag{42}$$

is smaller than unity, where M is the overall current mirror ratio from transistors T_f to the tail current of pair T_g. Using relations (21), (29) and (37), the adaptation law can then be expressed as

$$\frac{d\,c_{ij}}{d\,t} = a\ \sinh(s_i)\ \tanh(s_j/2) \tag{43}$$

where

$$a = \frac{2\,M\,I_B}{C\left(V_R - V_T\right)} \tag{44}$$

Condition (41) for this mode of operation could be enforced by choosing a very small fixed bias current I_B. The minimum value of I_B is however limited by the presence of leakage currents. Furthermore, values of S_i much larger than nU_T may be required to keep all offset voltages negligible. Since sinh(x) is a very steep function, large values of S_i will necessarily push transistors T_f into strong inversion. The mode of operation of T_g will then depend on the ratio R defined in (42).

f and g in strong inversion

Assuming that I_B becomes negligible, relations (21), (31) to (33) and (39) yield:

$$\frac{d\,c_{ij}}{d\,t} = \frac{a}{b}\sqrt{2bs_i^2 - s_j^2}\ \ s_j\ \ \mathrm{sgn}(s_i) \qquad \text{for } s_j^2 \le bs_i^2 \tag{45}$$

$$\frac{d\,c_{ij}}{d\,t} = a\ s_i^2\ \ \mathrm{sgn}(s_i\,s_j) \qquad \text{for } s_j^2 \ge bs_i^2 \tag{46}$$

where

$$a = \frac{K\,M\,I_{sf}}{C\left(V_R - V_T\right)} \tag{47}$$

and

$$b = \frac{4\,K\,M\,I_{sf}}{I_{sg}} \tag{48}$$

This adaptation law is represented in Fig.20. It cannot be decomposed into the product of two functions of s_i and s_j except for extreme values of factor b.

If *b>>1* then (45) is valid for most of the values of s_i and s_j (pair T_g almost never saturated) and can be approximated by

$$\frac{d\,c_{ij}}{d\,t} = \frac{a}{\sqrt{b/2}}\ s_i\ s_j \tag{49}$$

Functions f and g are both linear, which is not sufficient to reach the solution, as shown previously.

If $b<<1$, (46) is valid for most values of s_i and s_j. Differential pair T_g is almost always saturated and therfore behaves as a sign function. The two functions are then:

$$f(s_i) = \text{sgn}(s_i)\, s_i^2 \quad , \quad g(s_j) = \text{sgn}(s_j) \tag{50}$$

It must be pointed out that b <<1 implies R <<1. Therefore, the current density in transistors T_g is very low and they will tend to operate in weak inversion.

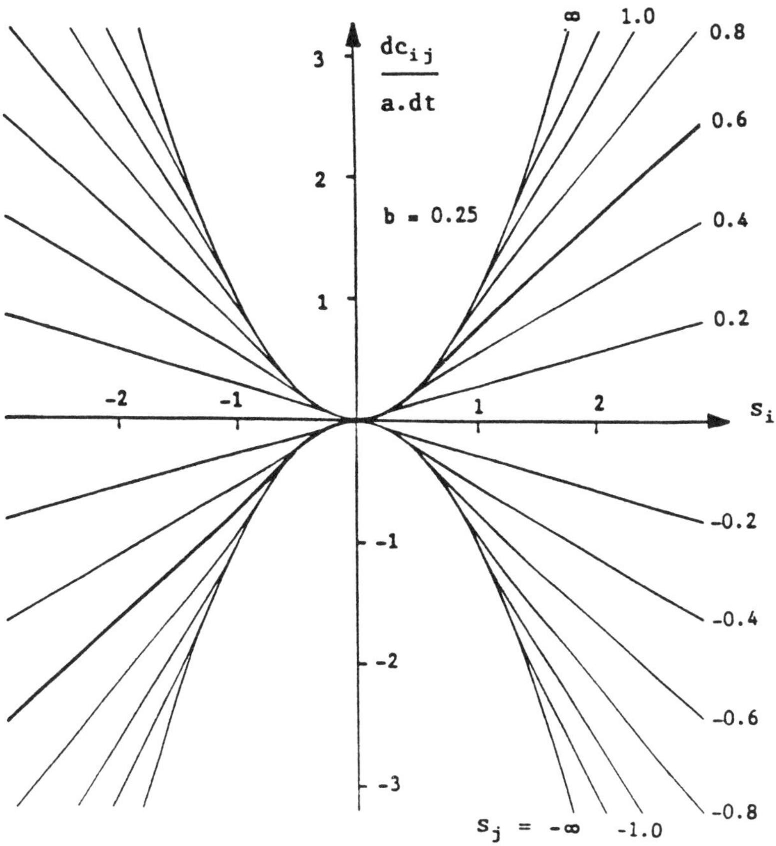

Fig.20 Adaptation law for f and g in strong inversion.

f in strong inversion and g in weak inversion

The adaptation law in this most important case is obtained by using relations (21), (29) and (39):

$$\frac{d\,c_{ij}}{d\,t} = a\,\mathrm{sgn}(s_j)\,s_i^2\,\tanh(s_j/2) \tag{51}$$

with gain a given by (47). This adaptation law is very close to that corresponding to (50), since the tanh(x) function for large values of x is equivalent to the sign function. It provides an excellent convergence behaviour, as shown by the simulation results in Fig.21. This figure also shows the time derivative of weight c_{12} as a function of output signal S_1, after equilibrium is reached. Its average value is indeed zero.

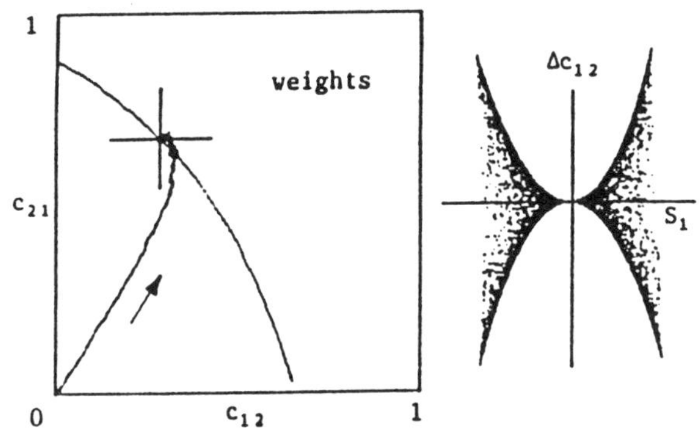

Fig.21 Convergence with T_f in strong inversion and T_g in weak inversion. The time derivative of c_{12} at equilibrium is also represented.

Fig. 22 to 24 show how this time derivative at equilibrium is affected by offset. It can be seen again that only one symmetrical function is needed to reach the correct solution.

In practical circuits, the maximum current, thus the maximum value of relation (46), will finally be limited by some transistors leaving saturation because of the limited power supply voltage.

As shown in the circuit diagram of Fig.19, provision has been made to modify the function f in the experimental circuit by adding a symmetrical offset voltage $\pm\,\Delta V_f$.

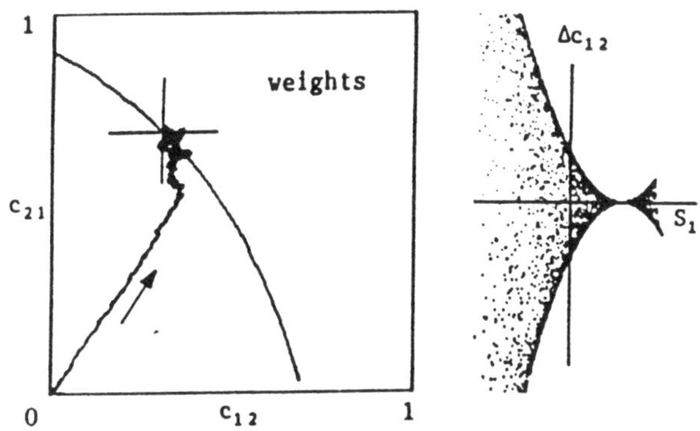

Fig.22 Convergence and time derivative of c_{12} at equilibrium with an offset in f equal to the RMS value of the sources.

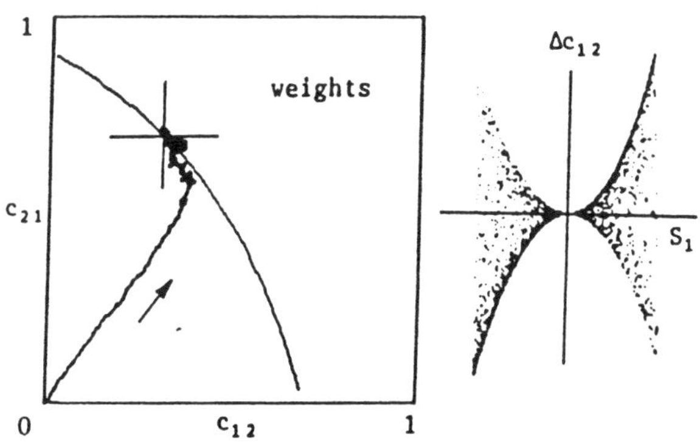

Fig.23 Convergence and time derivative of c_{12} at equilibrium with an offset in g equal to the RMS value of the sources.

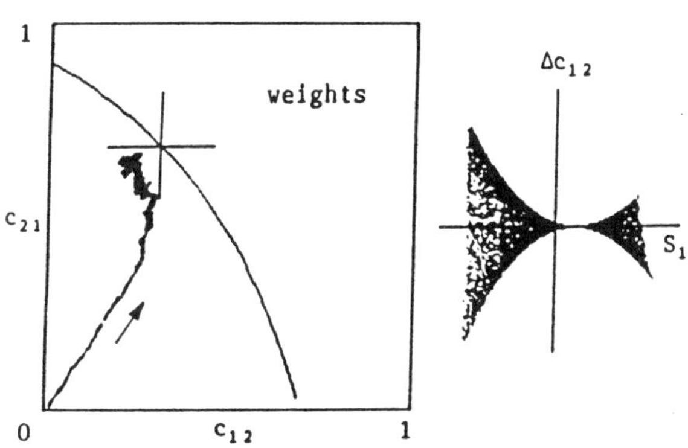

Fig 24 Convergence and time derivative of c_{12} at equilibrium when both functions are offset by 40% of the RMS value of the sources.

The complete block diagram of the experimental cell is shown in Fig.25. The f-function generator split into the negative and positve parts f^- and f^+ is shared by all blocks B, which have separate g-function generators. Signals S_j from 5 other cells can be connected, which allows a 6-cell, 30-synaptic weights network. Separate pins are provided to connect signals S_j to weighting transistor T_{ij} and to the g-function generator. This makes it possible to filter out the DC components before driving the latter. The network will then be insensitive to the offset of the amplifier OA or to any DC component in the input signals. To allow more flexibility, capacitors C that store the synaptic weights are not integrated. They are connected to separate pins, which also allow to measure the synaptic voltages V_{ij}.

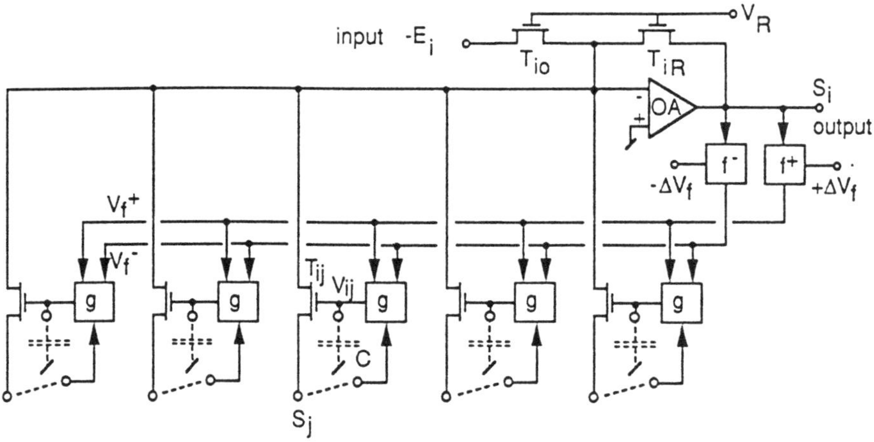

Fig.25 Complete diagram of the cell.

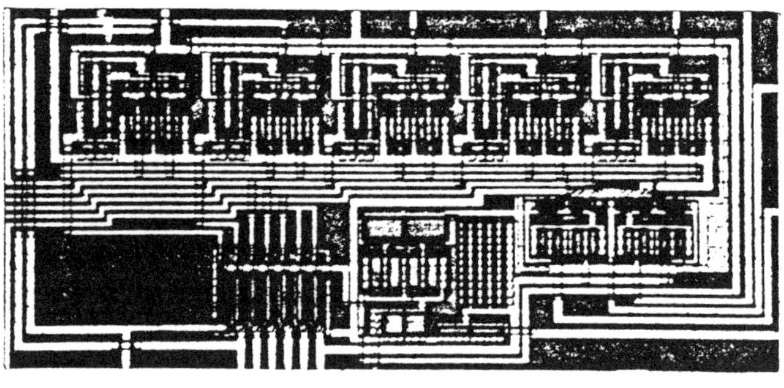

Fig.26 Microphotograph of the chip.

The whole circuit contains only 91 transistors. Thus a network of 6 cells will need a little more than 500 transistors and 30 capacitors. The cell has been integrated in a 3 μm silicon gate process [9]. Figure 26 is a microphotograph of the chip which has an area of about 0.36 mm^2 excluding the pads.

EXPERIMENTAL RESULTS

In this early design, a value of about 70 was errouneously selected for coefficient b (relation (48)). As shown by equation (49), this high value of b is equivalent to having linear f and g functions, which is not sufficient to force the output signals to be independent (solution P). For this reason a symmetrical offset voltage $\pm \Delta V_f$ of 50 mV has been applied to the f-function generator.

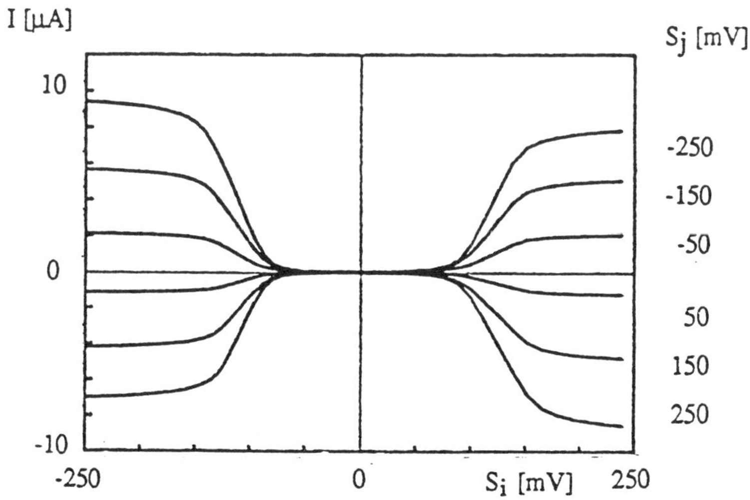

Fig.27 Measured output current of block B in short circuit.

Figure 27 shows the output current of one adaptation block B, measured as a function of S_i for various values of S_j. Except for the fact that they saturate (limited supply voltage), these curves are still somewhat comparable to those of Fig.20.

Figures 28 to 30 show experimental results obtained with a 2-cell network for various kinds of sources X_1 and X_2. The synaptic weights c_{12} and c_{21} are first set to zero by short-circuiting the capacitors. As a result, output signals S_1 and S_2 are a mixture of the sources, identical to those imposed as input E_1 and E_2. The short circuit is then suppressed (shown by an arrow) and the synaptic weights are let reach equilibrium. At equilibrium, the network clearly separates the sources, with a crosstalk of just a few percents. This is also the case when the sources are both approximately gaussian noises delivered by different generators. This is shown by the fact that the instantaneous differences S_1-X_1 and S_2-X_2 are drastically reduced at equilibrium. The explanation for such an unexpected and interesting result is not all clear [3].

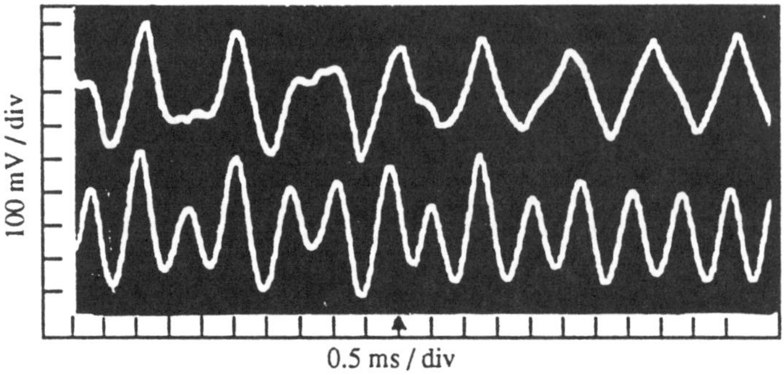

Fig.28 Experimental separation with a triangular and a sinusoidal source. $a_{11} = a_{22} = 1$, $a_{12} = 0.3$, $a_{21} = 0.7$

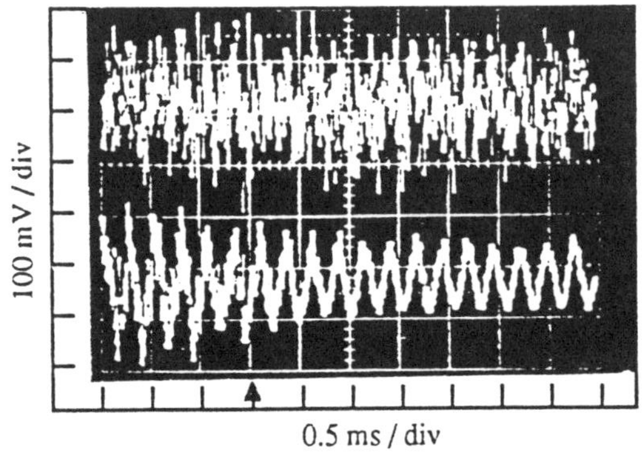

Fig.29 Experimental separation of a sinusoidal signal and a gaussian noise. $a_{11} = a_{22} = 1$, $a_{12} = 0.3$, $a_{21} = 0.7$

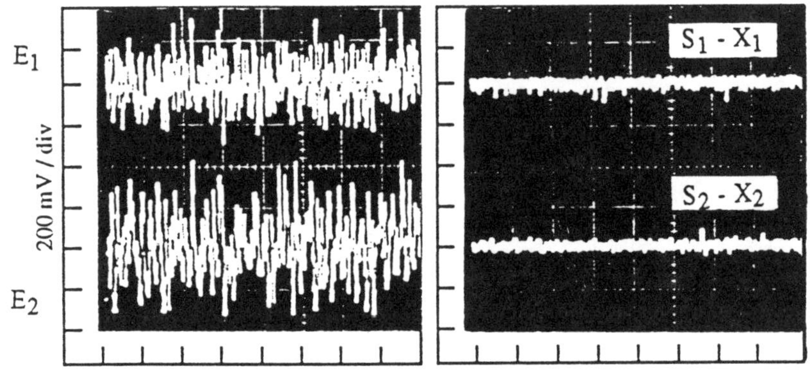

Fig 30 Experimental separation of two gaussian sources.

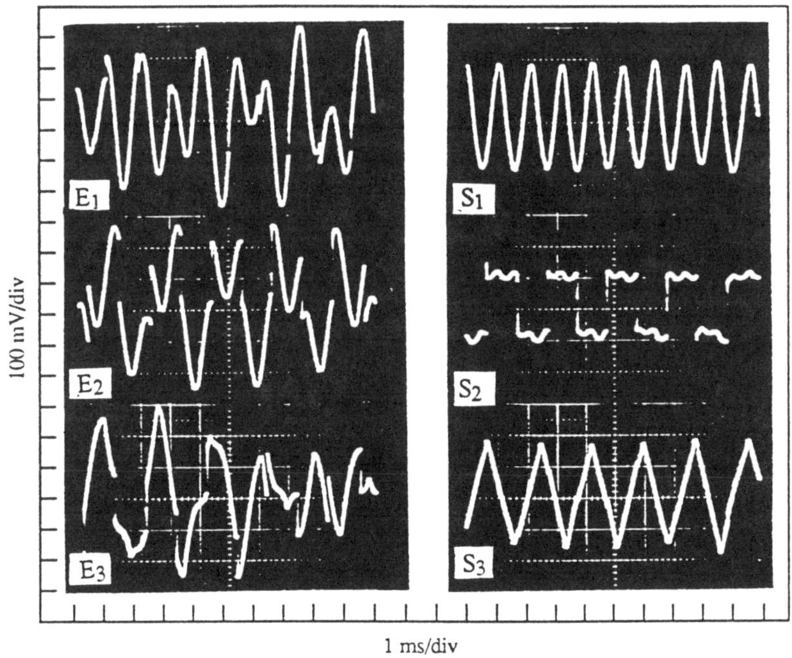

Fig.31 Experimental separation with a sinusoidal, a square and a triangular source. $a_{11}=a_{22}=a_{33}=1$, $a_{12}=a_{23}=a_{31}=0.3$, $a_{13}=a_{21}=a_{32}=0.7$.

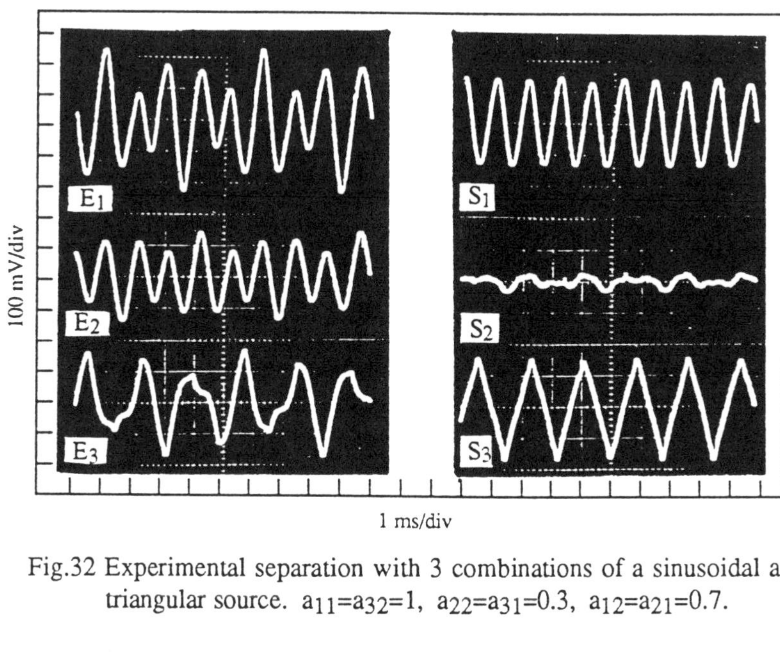

Fig.32 Experimental separation with 3 combinations of a sinusoidal and a triangular source. $a_{11}=a_{32}=1$, $a_{22}=a_{31}=0.3$, $a_{12}=a_{21}=0.7$.

Results for a 3-cell configuration are demonstrated in Figures 31 and 32. In Fig.32, only 2 independent sources are mixed to create 3 components of the input vector. The network finds the only solution for the output signals to be independent which is to extract the sources by 2 channels and put the third one to zero. The system is thus able to identify the number of independent sources which are combined in the input vector.

CONCLUSION

The algorithm proposed by Hérault and Jutten for separation of independent sources can be very efficiently implemented in a standard CMOS VLSI technology. The results obtained from a first experimental chip are very promising.

Besides an optimization of the present design, possible improvements include the introduction of some type of automatic gain control to allow a wider range of amplitudes. The weights might also be implemented in a more sophisticated way, to improve the maximum voltage acceptable without distortion. Bipolar operation [10] might be preferable to weak inversion for the pair of transistors T_g implementing the g-function generator, to reduce offset voltages. Integration of many cells in a single circuit will require to put the capacitors on chip. The current supplied by the adaptation block B will then have to be scaled down, as needed with low values of capacitors C. For very low frequency sources, solutions will have to be found to implement very low values of gain a, while keeping negligible the effect of leakage currents.

Many applications of this circuit can be expected [3]. Examples are identification of independent sources, separation of signal delivered by sensors, solution of the "cocktail party" problem in sonars, hearing aids or mother-foetus ECG measurements. Useful applications might possibly be found in image processing, including texture extraction.

AKNOWLEDGEMENT

The authors want to express their gratitude to C.Jutten and to E. Sorouchyari for very fruitful discussions.

REFERENCES

[1] J.Hérault, C.Jutten, "Space or time adaptive signal processing by neural network models", Neural networks for computing, Snowbird, 1986.

[2] C.Jutten, J.Hérault, A.Guerin, "IN.C.A: an INdependent Component Analyser based on an adaptive neuromimetic network", in "Artificial Intelligence and Cognitive Sciences", J.Demongeot, T.Hervé, V.Riallé and C.Roche (Editors), Manchester Press, 1988.

[3] C.Jutten, "Calcul neuromimétique et traitement du signal, analyse en composantes indépendantes", Thèse de Doctorat d'Etat ès sciences physiques, INPG, Grenoble, 1987 (in French).

[4] C.Mead, "Analog VLSI and Neural Systems", Addison-Wesley Publishing Co., Reading, Mass., 1989.

[5] E.Vittoz, "The design of high-performance analog circuits on digital CMOS chips", IEEE J.Solid-State Circuits, vol.SC-20, pp.657-665, June 1985.

[6] Y.P.Tsividis, "operation and modeling of the MOS transistor", McGraw-Hill, New York, 1987.

[7] H.Oguey, S.Cserveny, "MOSFET modelling from low to high current density", Summer Course on Process and Device Modelling, ESAT, Leuven, 1983.

[8] M.Degrauwe, J.Rijmenants, E.Vittoz, H.De Man, "Adaptive biasing CMOS amplifiers", IEEE J.Solid-State Circuits, vol.SC-17, pp.522-528, June 1982.

[9] R.Luescher, J.S De Saldivar, "A high density CMOS process",ISSCC Dig.Tech.Papers (New York),1985, pp.260-261.

[10] E.Vittoz, "MOS transistors operated in the lateral bipolar mode and their application in CMOS technology", IEEE J.Solid-State Circuits, vol.SC-18, pp.273-279, June 1983.

4

CIRCUIT MODELS OF SENSORY TRANSDUCTION IN THE COCHLEA

John Lazzaro and Carver Mead
Department of Computer Science
California Institute of Technology
Pasadena, California, 91125

Nonlinear signal processing is an integral part of sensory transduction in the nervous system. Sensory inputs are analog, continuous-time signals with a large dynamic range, whereas central neurons encode information with limited dynamic range and temporal specificity, using fixed-width, fixed-height pulses. Sensory transduction uses nonlinear signal processing to reduce real-world input to a neural representation, with a minimal loss of information.

An excellent example of nonlinear processing in sensory transduction occurs in the cochlea, the organ that converts the sound energy present at the eardrum into the first neural representation of the auditory system, the auditory nerve. Humans can process sound input over a 120-dB dynamic range, yet the firing rate of an auditory-nerve fiber can encode only about 25 dB of sound intensity. Humans can sense binaural time differences of the order of ten microseconds, yet an auditory-nerve fiber can fire at most once per millisecond. Using limited neural resources, the cochlea creates a representation that preserves the information essential for sound localization and understanding. Moreover, this neural code expresses auditory information in a way that facilitates feature extraction by higher neural structures.

We are building silicon integrated circuits that model sensory transduction in the cochlea, both to explore the general computational principles of the cochlea, and to create potentially useful devices for sound understanding, for sound localization, and for cochlear prostheses. In this paper, we describe the architecture and operation of an integrated circuit that models, to a limited degree, the evoked responses of the auditory nerve. The chip receives as input a time-varying voltage corresponding to sound input, and computes outputs that correspond to the responses of individual auditory-nerve fibers. The chip models the structure as well as the function of the cochlea; all subcircuits in the chip have anatomical correlates. The chip computes all outputs in real time, using analog continuous-time processing.

NEURAL ARCHITECTURE OF THE COCHLEA

Both mechanical and electrical processing occur in biological cochleas. The sound energy present at the eardrum is coupled into a mechanical traveling-wave structure, the basilar membrane, which converts time-domain information into spatially encoded information by spreading out signals in space according to their time scale (or frequency). Over much of its length, the velocity of propagation along the basilar membrane decreases exponentially with distance. The structure also contains active electromechanical elements; outer hair cells have motile properties, acting to reduce the damping of the passive basilar membrane and thus allowing weaker signals to be heard. Axons from higher brain centers innervate the outer hair cells; these centers may dynamically vary the local damping of the cochlea, providing frequency-specific automatic gain control (Kim, 1984).

Inner hair cells occur at regular intervals along the basilar membrane. Each inner hair cell acts as an electromechanical transducer, converting basilar-membrane vibration into a graded electrical signal. Several signal-processing operations occur during transduction. Inner hair cells half-wave rectify the mechanical signal, responding to motion in only one direction. Inner hair cells primarily respond to the velocity of basilar-membrane motion, implicitly computing the time derivative of basilar-membrane displacement (Dallos, 1985). Inner hair cells also compress the mechanical signal nonlinearly, reducing a large range of input sound intensities to a manageable excursion of signal level.

Spiral-ganglion neurons connect to each inner hair cell, and produce fixed-width, fixed-height pulses in response to inner-hair-cell electrical activity. The synaptic connection between the inner hair cell and the spiral-ganglion neuron may implement a stage of automatic gain control, exploiting the dynamics of synaptic-transmitter release (Geisler and Greenberg, 1986). Auditory-nerve fibers are axons from spiral-ganglion neurons; these fibers present a neural representation of audition to the brain.

When pure tones are presented as stimuli, an auditory-nerve fiber is most sensitive to tones of a specific frequency. This characteristic frequency corresponds to maximum basilar-membrane velocity at the location of the inner hair cell associated with the nerve fiber. The spiral trunk of the auditory nerve preserves this ordering; the nerve fibers are mapped cochleotopically and tonotopically. The mean firing rate of an auditory fiber encodes sound intensity, over about 25 dB of dynamic range. The temporal pattern of nerve firings reflects the shape of the filtered and rectified sound waveform; this phase locking does not diminish at high intensity levels (Evans, 1982).

SILICON MODELS OF THE COCHLEA

Both mechanical and electrical processing occur in biological cochleas. In the chip, however, we model both types of computation using electronic processing. A silicon model of the mechanical processing of the cochlea has been previously described (Lyon and Mead, 1988a; Mead, 1989). The circuit is a one-dimensional physical model of the traveling-wave structure formed by the basilar membrane. In this viewpoint of cochlear function, the exponentially tapered stiffness of the basilar membrane and the motility of the outer hair cells combine to produce a pseudoresonant structure.

The basilar-membrane circuit model implements this view of cochlear hydrodynamics using a cascade of second-order sections with exponentially scaled time constants. The cascade structure enforces unidirectionality, so a discretization in space does not introduce reflections that could cause instability in an active model. An analog, continuous-time circuit implementation of the model computes the pressure at selected discrete points along the basilar membrane in real time.

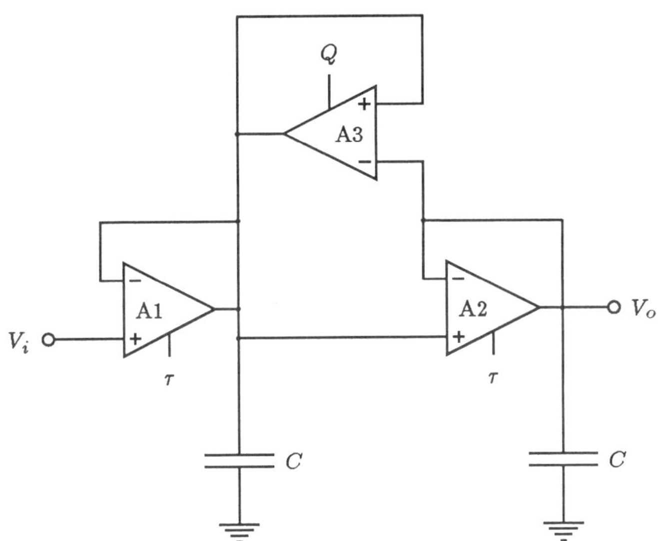

Figure 1. Circuit implementation of a second-order section. Input V_i and output V_o are time-varying voltages. The τ and Q control inputs set bias currents on transconductance amplifiers $A1$, $A2$, and $A3$, to control both the characteristic frequency and the peak height of the lowpass-filter response.

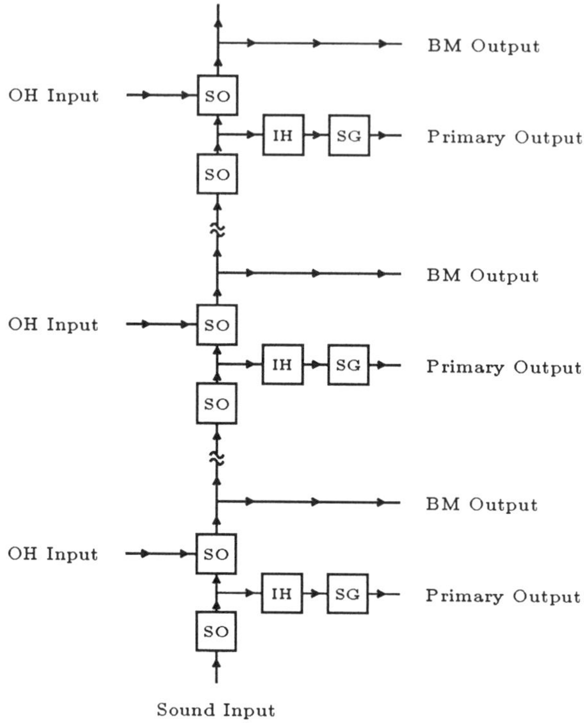

Figure 2. Block diagram of the chip. A time-varying input voltage, representing sound input to the cochlea, travels down the basilar-membrane model, a cascade of second-order sections (SO) with exponentially increasing time constants. Basilar-membrane (BM) circuit outputs show pressure along the membrane, whereas inputs modeling innervation of outer hair cells (OH) control local damping of the membrane circuit. Taps along the basilar membrane connect to a circuit model of inner hair cells (IH); outputs from inner hair cells connect to circuits that model spiral-ganglion neurons (SG). These neurons form the primary output of the chip, thus modeling auditory-fiber response.

Figure 1 shows the CMOS circuit implementation of a second-order section. Input and output signals for the circuit are time-varying voltages. The gain blocks are transconductance amplifiers, operated in the subthreshold regime. Capacitors are formed using the gate capacitance of n-channel and p-channel MOS transistors in parallel. Because of subthreshold amplifier operation, the time constant of the second-order section is an exponential function of the voltage applied to the transconductance control inputs of $A1$ and $A2$, labeled τ in Figure 1. Thus a cascade of second-order circuits, with a linear gradient applied to the τ control inputs, has exponentially scaled time constants. To implement this gradient, we used a polysilicon wire that travels along the length of circuit, and connects to the τ control input of each second-order section. A voltage difference across this wire, applied from off chip, produces exponentially scaled time constants. The amplifier $A3$ provides active positive feedback to the membrane, modeling the active mechanical feedback provided by the outer hair cells in biological cochleas. A second polysilicon wire is connected to the transconductance inputs of the $A3$ amplifiers in each second-order section (labeled Q in Figure 1); a voltage gradient across this wire similar to that on the τ control inputs sets all the second-order sections to the same response shape.

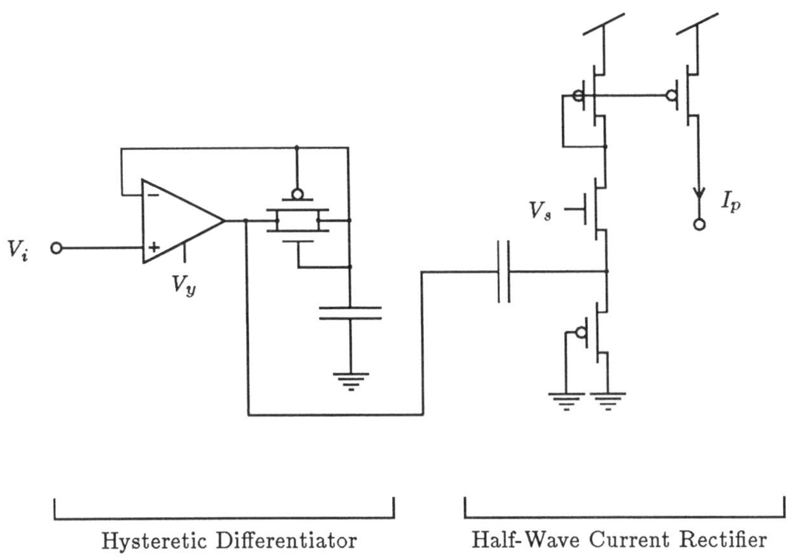

Figure 3. The inner-hair-cell circuit model. Input V_i, from the basilar-membrane circuit, is a time-varying voltage. The hysteretic-differentiator circuit, biased by voltage V_y, performs time differentiation and logarithmic compression. The output of the hysteretic differentiator, a time-varying voltage, connects to the half-wave current-rectifier circuit, which is shown in more detail in Figure 4.

This circuit model of cochlear mechanics is the foundation of our integrated circuit; Figure 2 shows the complete architecture of the chip. A way to model the adjustment of basilar-membrane damping by higher brain centers is to use an automatic-gain-control system that varies the damping of the second-order sections locally. We have not implemented this automatic-gain-control system; however, we have brought off chip several taps from the polysilicon wire that connects to the Q control of the second-order sections, allowing off-chip experiments with automatic gain control. To complete our circuit model of the auditory periphery, we have added circuits that model inner-hair-cell and spiral-ganglion-neuron functions

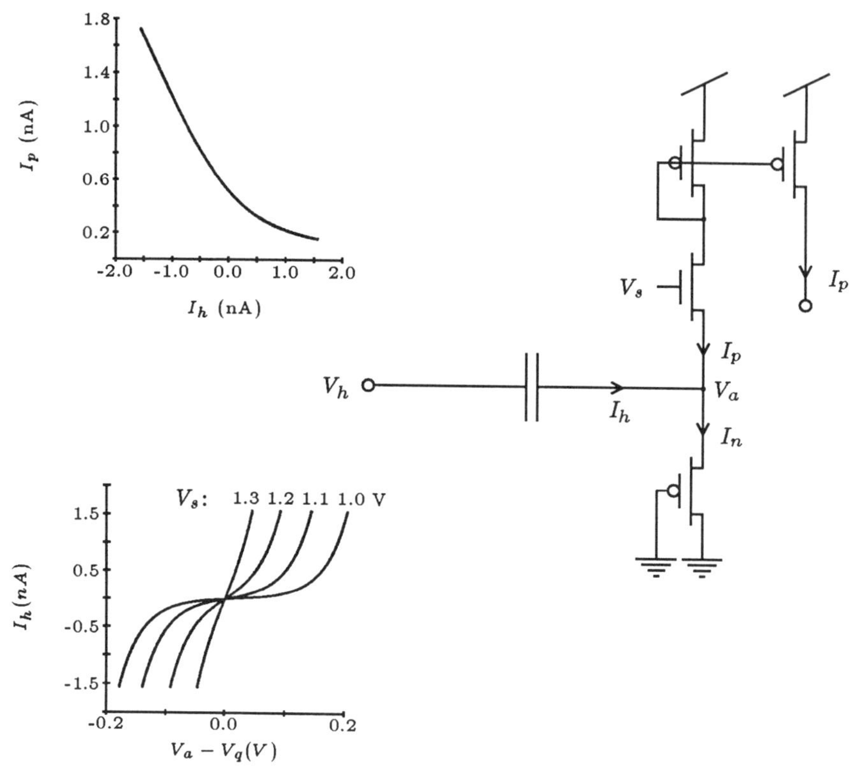

Figure 4. The half-wave current-rectifier circuit. Input V_h, from the hysteretic-differentiator circuit, is a time-varying voltage. A floating capacitor couples V_h into the node associated with V_a, as the bidirectional time-varying current I_h. The bottom graph shows the change in V_a required to sink or source I_h, for several values of bias voltage V_s; the voltage V_q is the value of V_a when $I_h = 0$. When $V_a = V_q$ and $I_h = 0$, the circuit output, the unidirectional current I_p, is at a quiescent value, I_q, set by V_s. Nonzero values of I_h modulate the output current I_p about I_q; for large $|I_h|$ relative to I_q, the circuit output I_p is a half-wave–rectified version of I_h, as shown in the top graph. Graphs show theoretical responses.

Figure 3 shows our inner-hair-cell circuit model. A hysteretic-differentiator circuit (Mead, 1989) processes the input-voltage waveform from the basilar-membrane circuit, performing time differentiation and logarithmic compression. The circuit enhances the zero-crossings of the input waveform, accentuating phase information in the signal. The output voltage of the hysteretic differentiator connects to a novel implementation of a half-wave current rectifier.

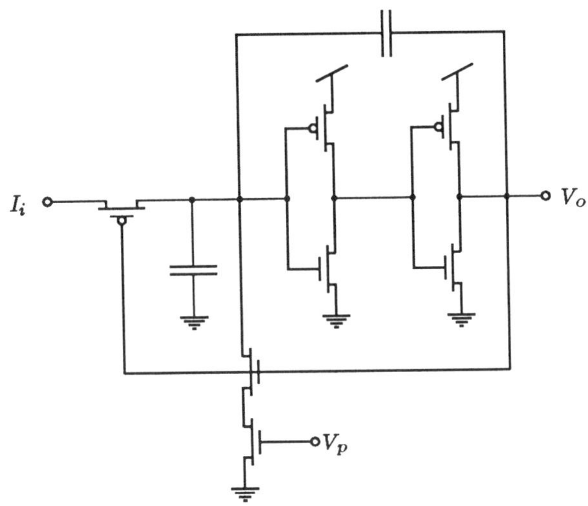

Figure 5. The spiral-ganglion-neuron circuit. Circuit input, from the half-wave–rectification circuit, is the unidirectional current I_i. The circuit converts this current into fixed-width, fixed-height voltage pulses, at output V_o. The bias voltage V_p sets pulse width; the output voltage V_o pulses between V_{dd} and ground.

Figure 4 shows our half-wave current-rectifier circuit. To understand its operation, consider the state of this circuit when the input voltage V_h is constant. If V_h is constant, $I_h = 0$, and V_a adapts such that $I_p = I_n$. For $I_h = 0$, we define the quiescent conditions $I_q \equiv I_p = I_n$ and $V_q \equiv V_a$. The value of I_q depends on the circuit bias voltage, V_s. A current mirror reflects this quiescent current to the circuit output. Thus, the output of the half-wave current-rectifier circuit in response to a constant voltage input is an adjustable bias current.

Now consider the circuit state when the input voltage V_h is a time-varying waveform. During the positive-going phase of the waveform, the current I_h is positive, and $I_n = I_h + I_p$. As I_n increases, V_a must also increase; the amount of increase depends on the circuit bias voltage, V_s, as shown in the bottom graph in Figure 4. However, if V_a increases, I_p must decrease. So, during the positive-going phase of the waveform, the output current I_p decreases from the quiescent current I_q.

During the negative-going phase of the waveform, the current I_h is negative, $I_p = |I_h| + I_n$, and the output current of the circuit increases from the quiescent current I_q. Thus, the circuit converts the input time-varying voltage waveform V_h into a unidirectional current waveform I_p. For large $|I_h|$ relative to I_q, the current waveform I_p is not symmetrical about I_q, and the average value of I_p is greater than that of I_q; thus, the circuit performs the rectification function, as shown in the top graph in Figure 4.

The current I_p is the output of the inner-hair-cell circuit. The spiral-ganglion neuron circuit model, shown in Figure 5, converts this current into fixed-width, fixed-height pulses. The circuit — a slightly modified version of the neuron circuit in (Mead, 1989) — creates a pulse rate that is linear in input current, for sufficiently low pulse rates. Thus, the average pulse rate of the circuit reflects the average value of I_p, whereas the temporal placement of each pulse reflects the shape of the current waveform I_p.

SILICON BASILAR-MEMBRANE RESPONSE

To test the tuning properties of the silicon auditory-nerve fibers, we duplicated a variety of classical auditory-nerve measurements. In these experiments, we tuned the basilar-membrane circuit to span about seven octaves, from 50 Hz to 10,000 Hz. We set the maximum firing rates of the auditory-fiber outputs at 150 to 300 spikes per second, with spike widths of 5 to 20 μs.

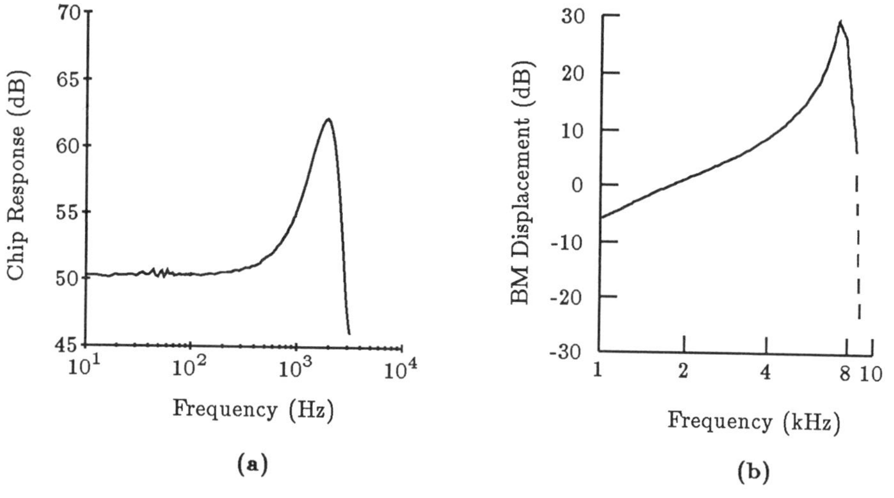

Figure 6. a: The response of the basilar-membrane circuit at a single point, to pure tones at a fixed input amplitude (0 dB = 3 mV peak). **b:** Transfer function of a single position on the basilar membrane of the squirrel monkey (Rhode, 1971). The curves show amplitude of vibration for constant malleus displacement.

In this configuration, without an input signal, the auditory-fiber outputs fire at less than 0.1 spike per second. At the characteristic frequency of a fiber, pure tones of a few millivolts peak amplitude produce responses significantly above this spontaneous rate. The chip can process tones up to about 1 V of peak amplitude, yielding approximately 60 dB of usable dynamic range.

Adding a preprocessor to basilar-membrane circuit, to limit intense input signals, would extend the upper limit of the dynamic range. A biological cochlea has a mechanical limiter as a preprocessor — the stapedial reflex. Designing more sensitive inner-hair-cell circuits would extend the lower limit of dynamic range. Both dynamic-range enhancements are currently under development.

Figure 6(a) shows a frequency-response plot for the basilar-membrane circuit, at a position with a best frequency of about 1900 Hz. The plot shows a flat response for frequencies significantly below the best frequency, a 12-dB response peak at the best frequency, and a sharp dropoff to the noise floor for frequencies significantly above the best frequency. This response is qualitatively similar to the frequency-response curve taken from the basilar membrane of the squirrel monkey using the Mossbauer effect, shown in Figure 6(b) (Rhode, 1971). Near the best frequency, basilar-membrane pressure, computed by the chip, is approximately equal to basilar-membrane displacement, measured by Rhode. Quantitatively, the bandwidth of the resonance peak of the chip response is wider than that of the physiological data; a cascade of second-order sections does not yield an optimal model of cochlear hydrodynamics (Lyon and Mead, 1988b).

The resonance peak of the chip response decreases for large-amplitude sinusoids, because the feedback amplifier $A3$ in the second-order sections saturates. The resonance peak in a physiological cochlea also decreases for large-amplitude inputs (Rhode, 1971). The silicon and physiological cochleas may show decreased resonance for similar reasons; for high sound intensities, outer hair cells in the physiological cochlea may not be capable of a linear response to basilar-membrane motion. Alternatively, an automatic-gain-control system may increase basilar-membrane damping locally for high-intensity sounds, by modulating the mechanical effect of the outer hair cells (Kim, 1984).

TUNING PROPERTIES OF THE SILICON AUDITORY NERVE

We characterized the tuning properties of the auditory-nerve-fiber circuit model, using pure tones as input. In response to a pure tone of sufficient intensity and appropriate frequency, the silicon auditory fiber produces spikes at a constant mean rate, as shown in Figure 7. The mean spike rate of a silicon fiber, in response to a constant tone, does not decrease over time, unlike that of a physiological auditory fiber; this lack of adaptation indicates the absence of dynamic automatic gain control in our model.

Figure 8(a) shows the mean spike rate of a silicon auditory fiber as a function of pure tone frequency. For low-amplitude tones, the fiber responds to a narrow range of frequencies; for higher-intensity tones, the fiber responds to a wider range of frequencies. The saturating nonlinearities of the basilar-membrane circuit and of the inner-hair-cell circuit cause the bandwidth of the fiber to increase with sound intensity. Qualitatively, this behavior matches the iso-intensity plots from an auditory-nerve fiber in the squirrel monkey (Rose et al., 1971), shown in Figure 8(b). Quantitatively, the saturation of the amplifiers in the forward path ($A1$ and $A2$) produce a detuning that is not a proper model of basilar-membrane mechanics.

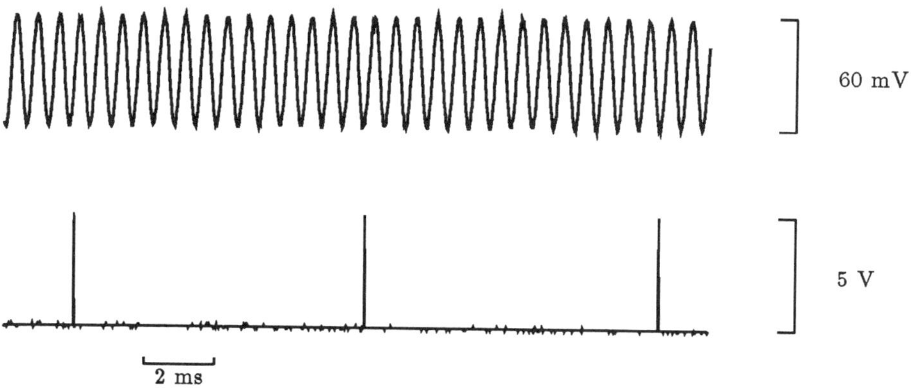

Figure 7. Output of a silicon auditory fiber (bottom trace) in response to a sinusoidal input (top trace). The frequency of the input is the characteristic frequency of the fiber.

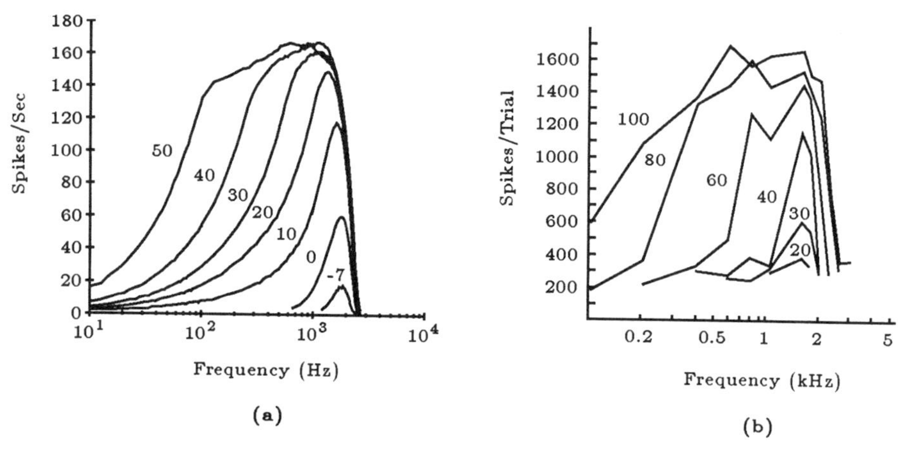

Figure 8. a: Plots showing the mean spike rate of a silicon auditory fiber as a function of pure tone frequency. Legend numbers indicate tone amplitude, in dB. **b:** Plots showing the number of discharges of an auditory fiber in the squirrel monkey, in response to a 10-s pure tone (Rose et al., 1971). Legend numbers indicate tone amplitude, in dB.

Figure 9(a) shows the mean spike rate of a silicon auditory fiber as a function of pure tone amplitude, at frequencies below, at, and above the best frequency of the fiber. In response to its characteristic frequency, 2100 Hz, the fiber encodes about 25 dB of tone amplitude before saturation. Figure 9(b) shows rate-intensity curves from an auditory fiber in the cat (Sachs and Abbas, 1974). At its characteristic frequency, the physiological fiber also encodes about 25 dB of tone amplitude before saturation. The shape of the biological and silicon curves at the characteristic frequency is remarkably similar, giving us some confidence in the validity of this modeling paradigm. In response to frequencies below and above the characteristic frequency, the functional forms of the silicon fiber responses are different from those of the physiological data. Most notably, the saturation rate of a silicon fiber for frequencies below the fiber's characteristic frequency exceeds the saturation rate of the silicon fiber at the fiber's characteristic frequency. This behavior is also a direct result of the undesired saturation at high input intensities of second-order-section amplifiers A1 and A2, shown in Figure 1, which model the stiffness of the basilar membrane. Above its best frequency, the response of the model decreases in a manner that is reminiscent of its biological counterpart.

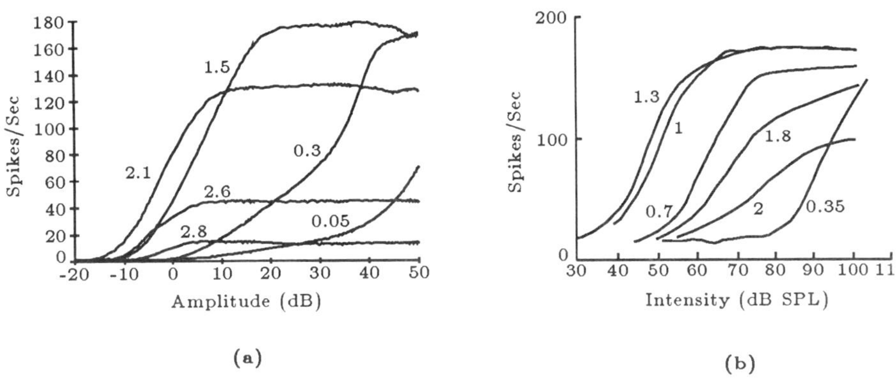

Figure 9. a: Plots showing the mean spike rate of a silicon auditory fiber as a function of pure tone amplitude. Legend numbers indicate tone frequency, in Hz. **b:** Plots showing the mean spike rate of an auditory fiber in the cat, as a function of pure tone amplitude (Sachs and Abbas, 1974). Legend numbers indicate tone frequency, in Hz.

Figure 10(a) shows iso-response curves for four silicon auditory-nerve fibers These plots represent an iso-rate section through the iso-intensity curves of Figure 8(a), at a spike rate for each fiber that was comfortably above the spontaneous rate. The chip response accurately models the steep high-frequency tail of tuning curves from cat auditory fibers (Kiang, 1980), shown in Figure 10(b); the shapes of physiological and chip tuning curves are qualitatively similar.

The bandwidth of the chip fibers for low sound intensities, however, is significantly wider than that of the physiological response. This problem stems from the wider bandwidth of the basilar-membrane circuit model, relative to that of the physiological data, as well as from the lack of a dynamic automatic-gain-control system for modulating the damping of the basilar-membrane circuit. The high-frequency cutoff of the iso-response curves, shown in Figure 10(a), is much steeper than is the cutoff of the iso-input curves shown in Figure 8(a). In a linear system, these two measurements would give identical results. The difference reflects the presence of a saturating nonlinearity in the system; the inner-hair-cell circuit and the basilar-membrane circuit provide this saturation function.

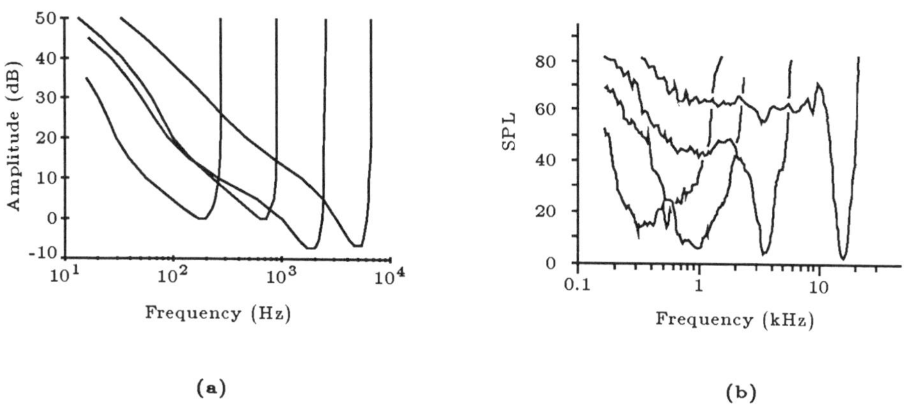

Figure 10. **a:** Plots showing iso-response curves for four silicon auditory fibers. The plots represent an iso-rate section through the iso-intensity curves of each fiber. Constant rates for each curve are, from the highest-frequency curve downward, 21.5, 16, 61, 59 spikes/s. **b:** Plots showing tuning curves from auditory fibers in the cat (Kiang, 1980). Fifty-ms tone bursts were presented at 10/s. Each tuning curve shows the sound pressure level (SPL) at the tympanic membrane (eardrum) that generates 10 spikes/s more activity during the tone bursts than during the silent interval.

TIMING PROPERTIES OF THE SILICON AUDITORY NERVE

The temporal firing patterns of the silicon auditory-nerve fibers encode information. Figure 11(a) shows period histograms of a chip fiber, in response to −5- to 50-dB pure tones at the fiber's characteristic frequency (0 dB = 3 mV peak); these histograms show the probability of a spike output occurring within a particular time interval during a single cycle of the input sinusoid. The fiber preserves the shape of the input sinusoid throughout this intensity range; this behavior matches data from an auditory fiber in the cat (Rose et al., 1971), shown in Figure 11(b). Unlike the cat fiber, however, the silicon fiber does not preserve absolute phase at higher intensities; this deficiency results from the saturation of the amplifiers $A1$ and $A2$ that model basilar-membrane stiffness. The temporal firing patterns of the silicon auditory-nerve fiber are, however, a good representation of signal periodicity; the synchronization ratios (normalized magnitude of the first Fourier coefficient) of the period histograms in Figure 11(a) are 0.5 to 0.6, comparable to those of physiological data at the same frequency (Rhode et al., 1978).

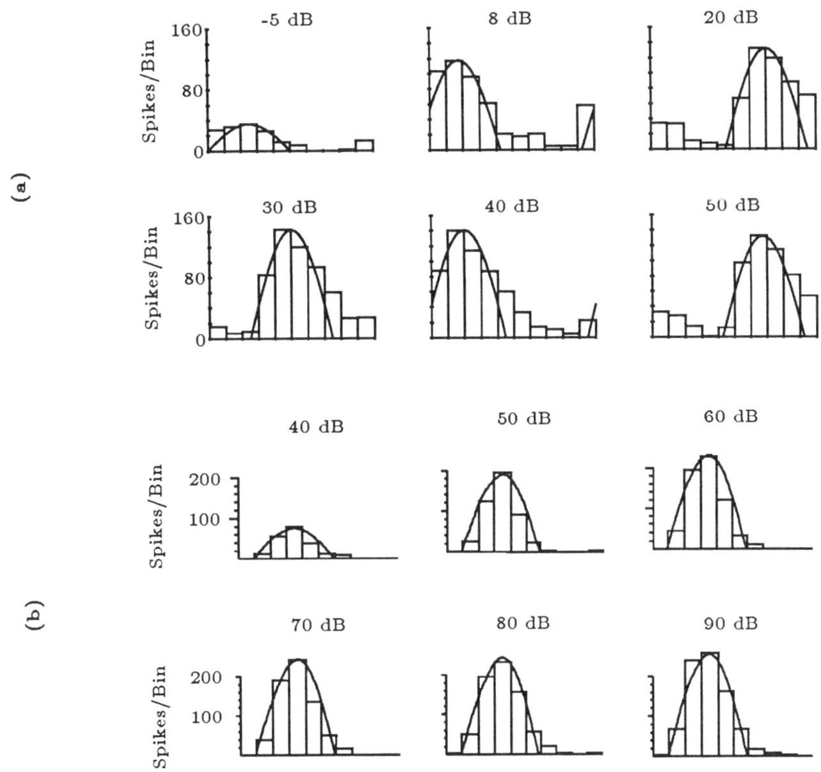

Figure 11. **a:** Period histograms of the silicon auditory-fiber response to a pure tone of 1840 Hz, near the fiber's best frequency. Amplitude of tone is shown above each plot. Histogram width is 54 μs. Each histogram begins at a constant position, relative to the input sinusoid; each is fitted to a sinusoid of best amplitude and phase. **b:** Period histograms of the response of an auditory fiber in the cat, to a low-frequency tone (Rose et al., 1971). Amplitude of pure tone is shown above each plot. Each histogram is fitted to a sinusoid of best amplitude but fixed phase.

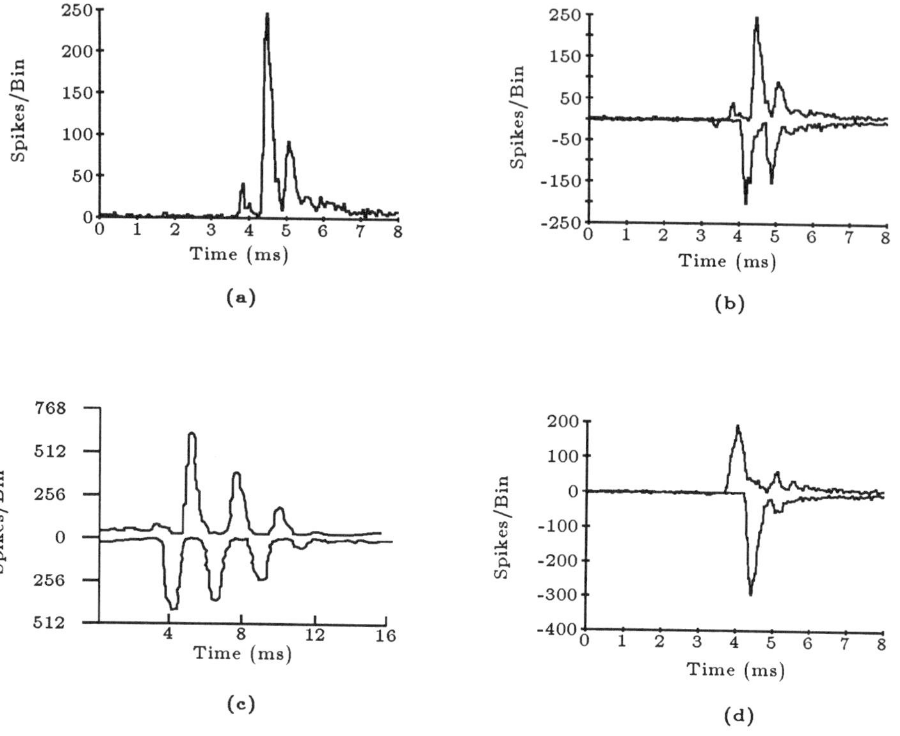

Figure 12. **a:** PST histogram of the rarefaction click response of a silicon auditory-nerve fiber. Click amplitude is 60 mV (26 dB peak); click width is 100 μs. Histogram is for 2000 click presentations; the width of each bin is 58 μs. **b:** Compound PST histogram of the click response of a silicon auditory-nerve fiber. Rarefaction click response is plotted as positive values; condensation click response is plotted as negative values. Conditions are identical to those of Figure 3(a). **c:** Compound PST histogram of the click response of an auditory fiber in the cat (Kiang et al., 1965). Click level is 30 dB relative to threshold response level; clicks width is 100 μs. Rarefaction click response is plotted as positive values; condensation click response is plotted as negative values. **d:** Compound PST histogram of the click response of a silicon auditory-nerve fiber, for a 200-mV click (36-dB click). All other conditions are identical to those of Figure 3(a).

The timing properties of silicon auditory-nerve fibers encode the click response of the basilar-membrane circuit. In response to a click of medium intensity, a silicon auditory-nerve fiber produces one or several spikes. To extract the click response from these spikes, we present the click stimulus to the chip many times, and record the responses of a silicon auditory-nerve fiber. These data are reduced to a poststimulus-time (PST) histogram, in which the height of each bin of the histogram indicates the number of spikes occurring within a particular time interval after the presentation of the click.

A PST histogram of the response of a silicon auditory-nerve fiber to a repetitive rarefaction click stimulus shows a half-wave–rectified version of a damped sinusoidal oscillation (Figure 12a). The frequency of this oscillation, 1724 Hz, is approximately the best frequency of the basilar-membrane position associated with this silicon nerve fiber. The half-wave rectification of the inner-hair-cell circuit removes the negative polarity of oscillatory waveform from the PST histogram of the click response. Repeating this experiment using a condensation click recovers the negative polarity of oscillation; a compound PST histogram, shown in Figure 12(b), combines data from both experiments to recreate the ringing waveform produced by the basilar-membrane circuit. Figure 12(c) shows a compound PST histogram of the click response of an auditory fiber in the cat (Kiang et al., 1965). Qualitatively, the circuit response matches the physiological response.

Figures 12(a) and 12(b) are chip responses to a 60-mV click stimulus (26 dB, 0 dB = 3 mV peak). Higher-intensity clicks produce oscillatory responses with increased damping; a compound PST histogram of chip auditory-nerve response to a 36-dB click shows reduced ringing (Figure 12d). This effect is a direct result of the nonlinear response of the basilar-membrane model; physiological basilar-membrane click responses also show reduced ringing at high click-intensity levels (Robles et al., 1976).

DISCUSSION

Our integrated circuit model captures many essential features of data representation in the auditory nerve; moreover, it computes the representation in real time. There are many traditional engineering representations of audition, however, that are also amenable to analog implementation. What advantages does a silicon auditory-nerve representation offer to a designer of artificial sensory systems?

As shown in Figures 11 and 12, an auditory-nerve fiber encodes a filtered, half-wave–rectified version of the input waveform, over a wide dynamic range, using the temporal patterning of fixed-width, fixed-height pulses. This representation supports the efficient, massively parallel computation of signal properties, using autocorrelations in time and cross-correlations between auditory fibers. In this representation, a correlation is simply a logical AND operation, performed by a few synapses in neural systems, or by a few transistors in silicon systems. Axonal delays in neural systems provide the time parameter for computing autocorrelations; in silicon systems, we model this delay with compact monostable circuits (Mead, 1989). We have used these techniques in a 220,000-transistor chip that models the auditory-localization system of the barn owl (Lazzaro and Mead, 1989).

The nonlinear filtering properties of the auditory-nerve fibers, shown in Figures 8 and 10, enhance these correlations. In a quiet environment, auditory fibers have narrow bandwidths; each fiber carries independent information, yielding rich correlations. In noisier environments, the tuning of auditory fibers widens, increasing the number of fibers that carry information about the signal. This detuning ensures that some fibers still encode signal properties reliably (Greenberg, 1988).

As shown in Figure 9, auditory fibers encode about 25 dB of signal intensity. Dynamic automatic gain control, present in a physiological cochlea, enhances this range; in addition, different populations of auditory fibers have different thresholds, further enhancing the encoding of signal intensity. Although not sufficient as a primary representation of sound, rate encoding of signal intensity is a valuable secondary cue, particularly for the detection of rapid spectral changes and the encoding of aperiodic sounds. Future versions of our chip will include these enhancements for rate encoding of signal intensity.

In conclusion, we have designed and tested an integrated circuit that computes, in real time, the evoked responses of auditory nerve, using analog, continuous-time processing. The chip offers a robust representation of audition, which can serve as a solid foundation for analog silicon systems that model higher auditory function.

Acknowledgements

We thank R. Lyon for valuable contributions throughout the project. We thank R. Lyon, M. Mahowald, L. Dupré, and D. Gillespie, for critically reading and correcting the manuscript. We thank Hewlett-Packard for computing support, and DARPA and MOSIS for chip fabrication. This work was sponsored by the Office of Naval Research and the System Development Foundation.

References

Dallos, P. (1985). Response characteristics of mammalian cochlear hair cells. *J. Neurosci.* **5**: 1591–1608.

Evans, E. F. (1982). Functional anatomy of the auditory system. In Barlow, H. B. and Mollon, J. D. (eds), *The Senses*. Cambridge, England: Cambridge University Press, p. 251.

Geisler, C.D. and Greenberg, S. (1986). A two-stage nonlinear cochlear model posseses automatic gain control. *J. Acoust. Soc. Am.* **80**: 1359–1363.

Greenberg, S. (1988). The ear as a speech analyzer. *J. Phonetics* **16**: 139–149.

Kiang, N. Y.-s, Watenabe, T., Thomas, E.C., and Clark, L.F. (1965). *Discharge Patterns of Single Fibers in the Cat's Auditory Nerve*. Cambridge, MA: M.I.T Press.

Kiang, N. Y.-s, (1980). Processing of speech by the auditory nervous system. *J. Acoust. Soc. Am.* **68**: 830–835.

Kim, D. O. (1984). Functional roles of the inner- and outer-haircell subsystems in the cochlea and brainstem. In Berlin, C. I. (ed), *Hearing Science*. San Diego, CA: College-Hill Press, p. 241.

Lazzaro, J. P. and Mead, C.A. (1989). Silicon models of auditory localization, *Neural Computation* **1**: 41–70.

Lyon, R. F. and Mead, C. A. (1988a). An analog electronic cochlea. *IEEE Trans. Acoust., Speech, Signal Processing* **36**: 1119–1134.

Lyon, R. F. and Mead, C. A. (1988b). *Cochlear Hydrodynamics Demystified.* Caltech Computer Science Technical Report Caltech–CS–TR–88–4, Pasadena, CA, February.

Mead, C. A. (1989). *Analog VLSI and Neural Systems*. Reading, MA: Addison-Wesley.

Rhode, W. S. (1971) Observations of the vibration of the basilar membrane in squirrel monkeys using the Mossbauer technique. *J. Acoust. Soc. Am.* **49**: 1218–1231.

Rhode, W. S., Geisler, C.D., and Kennedy, D.T. (1978). Auditory nerve fiber response to wide-band noise and tone combinations. *J. Neurophysiol.* **41**: 692–704.

Robles, L., Rhode, W. S., and Geisler, C.D. (1976) Transient response of basilar membrane measured in squirrel monkeys using the Mossbauer effect. *J. Acoust. Soc. Am.* **59**: 926–939.

Rose, J.E., Hind, J.E., Anderson, D. J., and Brugge, J. F. (1971). Some effects of stimulus intensity on response of auditory nerve fibers in the squirrel monkey. *J. Neurophysiol.* **34**: 685–699.

Sachs, M. B. and Abbas, P. J. (1974) Rate versus level functions for auditory-nerve fibers in cats: Tone-burst stimuli. *J. Acoust. Soc. Am.* **56**: 1835–1847.

5

Issues in Analog VLSI and MOS Techniques for Neural Computing

Steven Bibyk and Mohammed Ismail

Solid State Microelectronics Laboratory
Department of Electrical Engineering
Ohio State University, Columbus, Ohio 43210

Abstract

Issues in analog VLSI, such as the use of simple parameterized cells that are highly reconfigurable and input/output compatability, are being molded by the activities in developing hardware implementations of microelectronic neural networks. Analog MOS circuit modules, such as integrators, summers, and multipliers can be configured in a neural network architecture to build feedback/feedforward neural networks and/or the equivalent of adaptive, state-space signal processors. The methods of adaptation can be compared by evaluating a criterion or energy function which drives the adaptation process.

Introduction

Analog Very Large Scale Integration (VLSI) has recently been receiving considerable attention [1-4]. The motivation behind this is two-fold: 1) VLSI is now maturing with emphasis towards submicron structures and sophisticated applications combining digital as well as analog circuits on a single chip. Examples are found on today's advanced systems for telecommunications, robotics, automotive electronics, image processing, intelligent sensors, etc.; 2) Massive application of analog VLSI provides means for the hardware implementation of adaptive systems based on neural paradigms. As a result, analog VLSI has been recognized as a major technology for future information processing. To match such fast technological trends toward single chip mixed analog/digital VLSI systems, a wealthy activity on analog integrated circuits is underway and succesful attempts to review the state-of-the art in the analog field have recently been reported [5-8].

This chapter consists of three parts. First, we will discuss important issues in analog VLSI that are relevant to the hardware implementation of microelectronic neural networks in MOS technology. In the second part we will unveil interesting connections between neural paradigms, continuous-time adaptive systems, and analog signal processing. Novel and simple continuous-time analog MOS circuit techniques for the VLSI implementation of neural systems are described in the third part. The new circuit techniques are based on extremely simple and programmable analog parameterized cells with such attractive features as reconfigurability, input/output compatibility, and unrestricted fan-in/fan-out capability. The implementations presented take advantage of new continuous-time MOS circuit design concepts as well as advanced CMOS technologies.

Current Issues in Analog VLSI

Analog circuits in general, and integrated circuits in particular, are still designed largely by hand, by experts intimately familiar with nuances of the target application and integrated circuit fabrication process. Analog design is commonly perceived to be one of the most knowledge-intensive of design tasks. The techniques needed to build good analog circuits seem to exist solely as expertise invested in individual designers. In addition, the state of analog design tools is quite primitive in comparison to digital synthesis tools. Table 1 provides a general comparison [9] between analog and digital designs. Analog design "bottlenecks" have resulted in a lack of effective design tools, which has been the primary cause of major cycle time differences observed between analog and digital product development.

In recent years, however, the state of analog design tools has shown signs of dramatic changes. Design strategies and philosophies to bridge the gap between classical analog design and VLSI have been established. Inspired by the reaches of methodologies and techniques of digital VLSI, a large volume of activities on analog VLSI is currently underway. The goal of these efforts is to develop efficient tools for synthesis at both circuit and layout levels, simulation, and testing of large scale analog integrated circuits. The rate of progress of analog VLSI neural networks strongly depends upon the maturing of these efforts. Fortunately, some of the traditional analog design requirements such as accurate absolute component values, device matching, precise time constants, etc., are not major concerns in neural networks. This is primarily because computation precision of individual neurons does not seem to be of paramount importance. If analog tools are exclusively developed for neural network implementation, then these issues should be taken into account. It is also worth noting that some design factors such as dynamic range, signal handling, and frequency range are not well defined for neural networks and that the problems of interconnections and power consumption represent design hurdles.

Table 1: Analog vs. Digital Design

Analog Design	Digital Design
Signals have a continuum of values for amplitude and time	Signals have only two states
Irregular blocks	Regular blocks
Customized	Standardized
Components have a continuum of values	Components with fixed values
Requires precise modeling	Modeling can be simplified
Difficult to use with CAD	Amenable to CAD methodology
Designed at the circuit level	Designed at the system level
Longer design times	Short design times
Two-three tries necessary for success	Successful circuits the first time
Diffiicult to test	Amenable to design-for-test

Following a brief review of analog VLSI design methodologies, we will discuss important issues in analog tools from a neural hardware perspective.

Analog Design Methodologies

Several design methodologies [9] for analog integrated circuits (IC's) have been identified to represent the current and near future trends in the development of an analog VLSI system. These are:

Analog Arrays

Pre-processed IC's containing unconnected components and groups of connected components which are programmed by defining interconnections on one or more mask layers.

Analog Standard Cell/Block

Pre-designed circuits which reside in a software database and can be used to implement the design of an analog IC.

Analog Parameterized Cell/Block

Partially pre-designed circuits which reside in a sofware database and can be programmed or parameterized at the time of design of the IC.

Analog Programmable Chips

Completely fabricated chips which are capable of programming by electrical or some other means.

Analog Silicon Compilers

Automatic design (and layout) of analog circuits from a high level specification.

Since analog VLSI neural networks comprise a very large number of interconnected identical neurons they enjoy a very high degree of modularity. Therefore, the analog standard and parameterized cell design methodologies seem very appropriate. Electronic neural networks should make use of very simple building blocks with such features as reconfigurability, versatility and most importantly, simplicity. This results in a neural architecture that requires less design time and makes effective use of VLSI computer-aided design (CAD) tools [10]. The more reconfigurable/versatile the analog circuit is, the more it becomes like a digital cell. We advocate the design of primitives or well-defined analog cells that are input-output compatible and can be interconnected to achieve different linear and/or nonlinear functions. The design of the analog cells needs to be done in parallel with the design of neural processing modules, and the modes of neural processing should be matched to the characteristics of the analog cells. Such analog cells [10] will ultimately bring analog VLSI design in general, and neural networks in particular, a step closer towards automation. This brings us to some of the important tools for analog VLSI neural networks which will be discussed next.

Analog VLSI Design Tools

Here we discuss some of the important design tools which will significantly impact the rate of progress of analog VLSI neural networks and their applications both as synthetic elements for computational neuroscience and in the area of information processing.

Today's analog CAD design system, in general, is very much a "home brew" system composed of a collection of at best, loosely integrated proprietary, commercial, and university tools [11]. These are not highly automated but rather provide an environment which assists the analog designer. Fortunately, CAD-based analog synthesis is evolving and includes steps to map portions of the analog design into the digital realm to take advantage of existing digital VLSI CAD tools. This mapping is particularly useful for simulation above the circuit level and for chip-level physical design. Circuit simulation, e.g. SPICE or its derivatives, is the most mature tool. Analog schematic captures are also well-developed. However, poor integration of schematic capture and simulation and lack of support for partitioning and design space explorations remain to be solved.

Analog physical design makes use of symbolic layout techniques and of manual placement and automatic routing. Automatic module generation and techniques to bury critical nodes are used in order to apply digital tools to complete full-chip layouts. Some of the problems include inadequate handling of parasitics and inadequate post-layout verification. In general, design-for-test and design-for-manufacturability are not supported. Analog knowledge-based systems for the design of basic fixed cells such as the operational amplifier (op-amp) are now available [2,12,13]. The system represents circuit topologies as a hierarchy of functional blocks. A planning mechanism translates performance specifications between levels in this hierarchy. The system then provides schematic and layout of sized transistors for simple CMOS op-amps from performance specifications and process parameters. Circuit topologies as well as design equations are represented as statistically stored templates for use during the translation step.

Testing of Analog Neural Networks

Unlike testing of traditional analog IC's, the testing of electronic neural networks is a complex task involving the simultaneous control and acquisition of many analog signals. The number of possible test points grows rapidly with the number of neurons in a neural network integrated circuit. It is therefore impractical to test all possible nodes for large networks. Two types of testing emerge, each with distinct goals and requirements. These are: 1) Component test which involves characterization of individual components of a network, e.g. neural cells or synapses, with all test points taken to pins or probe-pads. This requires the input and output of multiple analog signals; 2) System test which involves functional characterization of the network using only system inputs and outputs. All nodes cannot be tested. Fortunately, the network should be tolerant of a certain number of non-functioning components. Also, since many of the signals will be adapted, component testing for meeting of design specifications is not as critical. Both types of testing involve simultaneous generation and observation of a large number of analog signals. The use of a computer controlled test is a necessity to manage testing complexity.

Neural Network Signal Processing

Shown in figure 1a is an analog circuit which utilizes adaptation in performing signal processing. The adaptation is represented as signals which will be called weights. The weight signals are integrator outputs, which can be thought of as determining pole and zero positions in a linear system. In general, when the system is nonlinear, the weight signals represent the key system time constants; the poles of the system are determined by adapting signal values. Only the time constants associated with the adaptation rate are determined by component values. Since these time constants do not have to be precise, they can be determined by the RC time constants of the weight integrator components. In many cases, the weight integrators can be treated as two input, single output circuits, where the output is the mutliplication of the weight signal and one of the inputs, as shown in figure 1b.

1A

1B

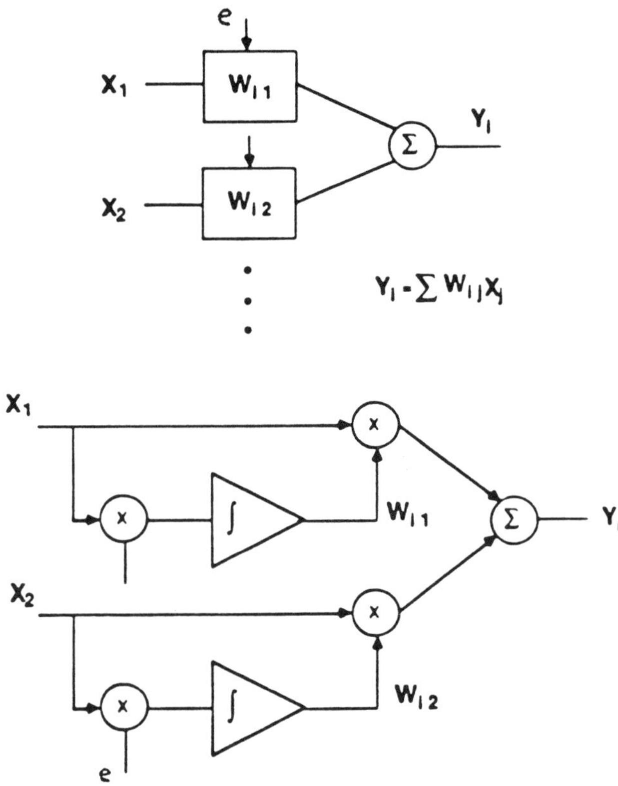

Figure 1: 1a. Block diagram of neuron and weight synapses. 1b. Synapses shown as integrators.

Many of the learning or adaptation rules of neural networks involve updating circuit signals called weights. In discrete time simulations of neural networks, the weight update is given by:

$$\omega_{ij}(k+1) = \omega_{ij}(k) + \text{ rule}(k)$$

where $rule(k)$ specifies the type of adaptation algorithm. The corresponding continuous time equation is:

$$\frac{\partial \omega_{ij}(t)}{\partial t} = \text{rule}(t)$$

Thus the weight signal can be taken at the output of an integrator whose input is the signal $rule(t)$. This signal takes on various forms depending on the type of learning rule.

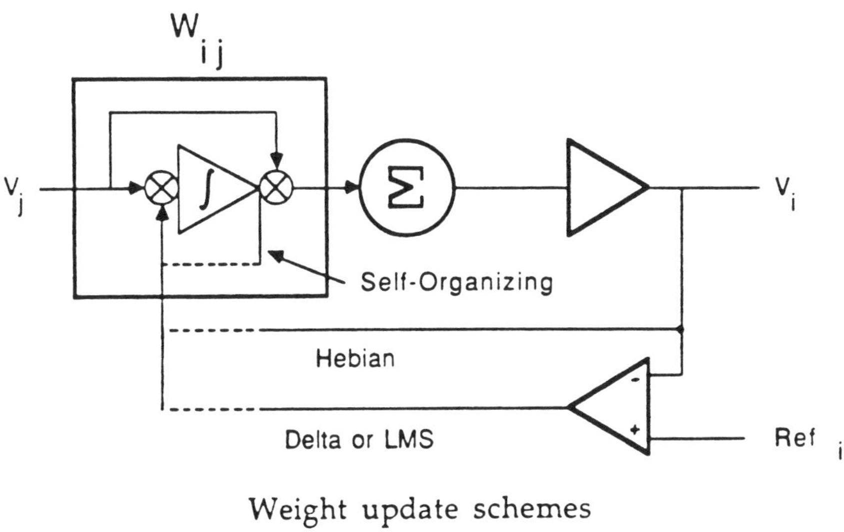

Figure 2: Circuit schematic showing weight adaptation for delta, hebbian, and self organizing learning.

Several types of learning rules are schematically shown for a single weight in figure 2. The circuit indicates the type of wiring and components that would be used for delta, Hebbian, and self organizing type learning. The weight is represented as a multiplier-integrator-multiplier circuit, whose inputs are a axon signal from another neuron plus an adaptation signal. The weights are combined with the axon signals V_j and amplified to produce the axon signal V_i. In some delta adaptation processes, a reference signal is compared to V_i to form an error signal. Different types of learning use alternate types of feedback connections into the weight integrators. By changing the wiring back into the integrators, we are changing the manner of adaptation. In general, in adaptive signal processing, a criterion is specified to drive the adaptation process. Some common criteria are minimizing a mean square error (MSE) or the establishment of a local reference signal. The criterion is usually specified as achieving the minumum or maximum of some function ξ of the weight signals and the input signals. At this extremum, the weights have their desired values, so it is useful to specify the time rate of change of the weights as given by the gradient in weight space of the criterion function, as described by:

$$\frac{\partial \omega_{ij}}{\partial t} = \pm \nabla_\omega \xi[\vec{\omega}(t),\ \vec{X}(t)]$$

where the vector $\vec{X}(t)$ is the input signal and the $\vec{\omega}(t)$ are the weights.

When the extremum is reached, the weights no longer change. Note that the criterion function for the weights serves the same role as the energy function given by Hopfield for describing the dynamics of the neural outputs when the weights are fixed [14]. In many cases, the criterion function is a bounded function in weight space, and the weights will change to either go up or down the gradient, depending on the system objective.

The circuitry which drives the weight integrators should naturally calculate the gradient of the criterion function. Different learning rules would use different circuit architectures. Unfortunately, the desired criterion is often too complicated to be calculated exactly. One typical example is the mean square error signal, which requires storing the square of the error signal for all time. This proves to be exceedingly difficult. However, given a desired criterion function, we are free to choose simpler functions which have extremums in weight space at the same locations as the original criterion. Since we are using gradients, the function is somewhat arbitrary, only the derivatives must be similar. A famous example is the use of the square of the instantaneous error, as well as its gradient, which is much simpler to compute [15] This is the LMS (Least Mean Square) rule or the generalized delta rule. The circuitry to compute the LMS gradient is quite simple, requiring a comparator and a multiplier, and no explicit memory. In many cases, the weights converge to the same value as they would have for the MSE criterion, although maybe not as quickly. This indicates a basic tradeoff between complexity of the adaptation rule vs. practicality of hardware construction. In this case, the gradient of the instantaneous error asymptotically goes to zero when the gradient of the MSE goes to zero, but the former is much easier to compute and construct in hardware.

Many of the learning rules which are being considered for hardware construction are simple by necessity, trying to take advantage of local wiring and minimal computational elements. In many cases, obtaining values from memory is an inefficient method of computation. One can expect that the learning rules are gradient estimates of a criterion function, and given the rule, it should be useful to consider what criterion is being driven to an extremum. Two rules which appear to be attractive for hardware implementation with analog circuitry are Hebbian learning and the self organizing networks. Both take advantage of local wiring to simplify the construction, yet can be quite powerful in signal processing capabilities. However, in some cases the learning rule is only an estimate, so that there is some arbitrariness in determining the desired criterion or energy function.

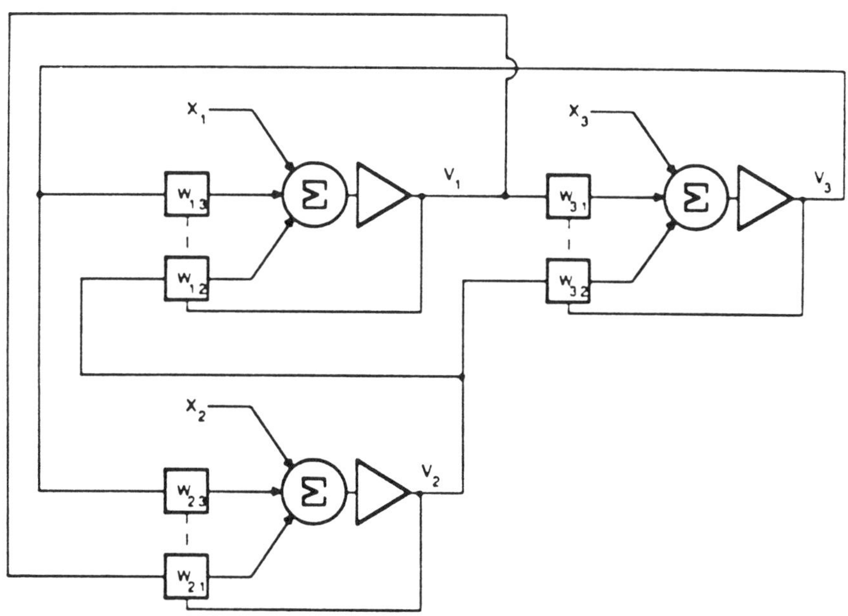

Figure 3: Novelty filter which uses Hebbian learning to adapt synapses.

A signal processing circuit which incorporates Hebbian adaptation is Kohonen's novelty filter [16], schematically indicated in figure 3. The Vs are the neural axon signals and the Xs are the input signals. Kohonen has analyzed this circuit for the linear case, and notes that for a given vector of input signals the neural outputs tend to go to zero. The hardware can be developed with saturating amplifiers (dual supply) and integrators, where the zero neural output voltage would correspond to a linear region of operation. However, the weight integrators will operate in their nonlinear region with saturating outputs. This has an advantage however, in that the poles of the system have a limited range, which can be used to insure stability.

The circuit illustrates the large amount of feedback which drives the neural outputs toward zero, independent of the input signals. In order to achieve this, the weight signals become a representation of the input signals over time. By increasing the number of neurons while keeping the number of inputs the same, we increase the precision of the weight representation, since the weight values collectively drive the neural outputs toward zero so that the input characteristics are spread out over more weights [17]. Since each weight is an integrator, this is a highly recursive system with a rich set of dynamics. The neural outputs are also state variables, whose dynamics are described by the Hopfield energy function when the weights are fixed. When both the neural outputs and the weights are state variables, the total energy function for the entire system is more complex. Nevertheless, there is an aesthetic appeal to the fact that since there is no reference signal, the neural outputs will drive toward zero, which represents a resting state.

Since the average neural output is zero, we can treat the actual neural output as representing the variance of a signal. Thus, a reasonable criterion for adaptation is that the weights are changing to minimize the sum of the squares of the neural outputs over time, which is a variance. Thus, the weights are the best representation of the input signals which minimize the variance of the neural outputs. The filter is very similar to the Kalman filter approach, except that we have a nonlinear estimate instead of a linear estimate. The type of nonlinearities are determined from the architecture of the neural circuit.

In order to develop the adaptation rule, the criterion can be simplified by using the instantaneous value of the neural outputs as opposed to the time averaged value. To take the gradient of the approximate criterion, suppose the criterion function ξ is given by:

$$\xi = \sum_i \int V_i^2(t)dt$$

$$V_i = f(\sum_j w_{ij} V_j \ + X_i)$$

Use the instantaneous gradient estimate

$$\nabla_w \sum_i V_i^2(t)$$

Expanding the gradient shows the recursive nature of the filter, as seen by

$$\frac{\partial \xi}{\partial \omega_{ij}} \approx \sum_k 2V_k \frac{\partial V_k}{\partial \omega_{ij}}$$

$$\frac{\partial V_k}{\partial \omega_{ij}} = \sum_m \omega_{km} \frac{\partial V_m}{\partial \omega_{ij}} + V_j$$

we use the approximation that $\partial V_k / \partial \omega_{ij} \approx V_j$, when $k = i$ and equals zero otherwise. Derivatives of the sigmoidal function f() were neglected. The above analysis is similar to the HARF (Hyperstable Adaptive Recursive Filter) approximation used in recursive LMS systems [15]. By changing weights using the negative of the gradient, we have

$$\frac{\partial \xi}{\partial \omega_{ij}} \approx uV_iV_j = -\frac{\partial \omega_{ij}}{\partial t}$$

which is a Hebbian learning rule. Thus, Hebbian learning corresponds to a gradient estimate which can minimize the average sum of squares of the neural outputs. Note that a plus sign (transversing up the gradient) for a discrete time system corresponds to

$$\omega_{ij} = \sum_s V_i^s V_j^s$$

which is the discrete form used by Hopfield to develop an associative memory.

In the Kohonen novelty filter, it is important that the input signals be multiplied by fixed weights, otherwise, these weights will adapt to zero and disconnect the effect of the inputs. The filter is equivalent to a recursive LMS filter where the reference signals are zero, and the input weights are not allowed to adapt. The recursive LMS filter is discussed in the following section.

Continuous-Time, Adaptive, Recursive Filtering

Shown in figure 4 is a filter which incorporates a type of adaptation that is equivalent to a recursive LMS, or a generalized delta which is also referred to as back propagation [18]. In this figure, there are two types of weights, those that have a local error signal (the w_{ij}), and those that do not (the S_{ij}). In nearly all cases, an internal reference signal is not available, and must be calculated from the error signals in the outermost layer. In figure 4, to calculate

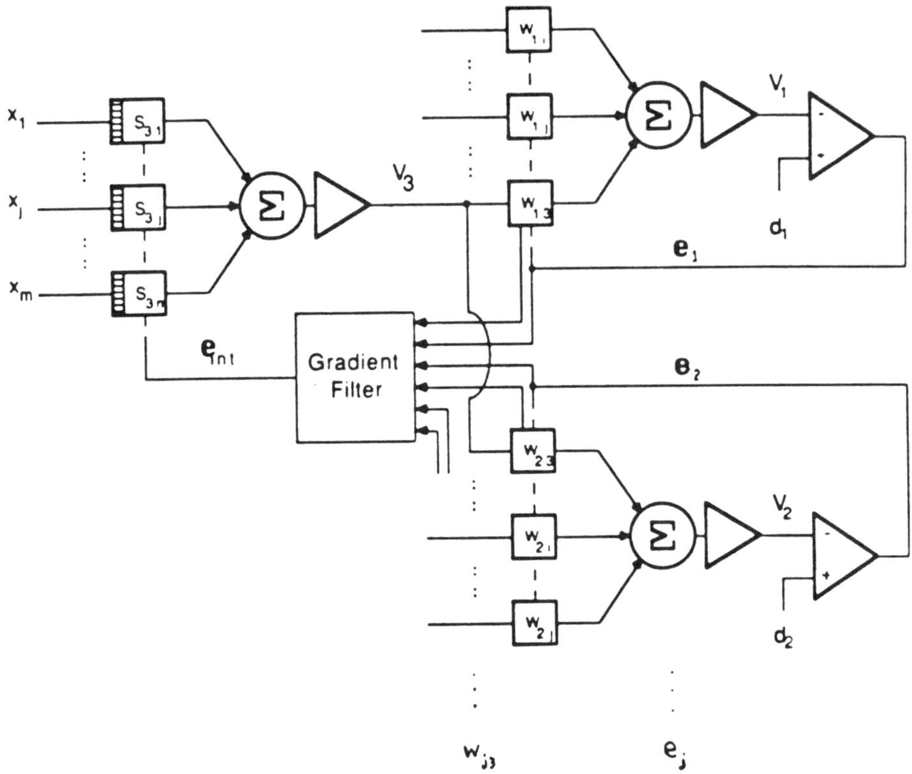

Figure 4: Neural circuit to be used for analog back propagation.

the internal error e_{int} for the weights S_{3j}, the outer layer errors e_j are multiplied by the transpose weights w_{j3}, to form the sum $\sum_j w_{j3}e_j$ which yields e_{int}. The circuit block which performs this function is called the gradient filter in figure 4. In neural simulations, a related method is known as the back propagation technique, although we have neglected the derivatives of the sigmoidal functions for clarity. To build the gradient filter requires use of a transpose filter of the weight integrators. A similar technique is used for recursive, adaptive, state space filters, an example of which is shown in figure 5, using state space notation. The c's correspond to weights which have a local error signal available, whereas the As correspond to weights which are internal and do not have a local error signal.

An interesting approach using sensitivity analysis for analog, state space recursive fiters is described by Johns et al. [19]. These filters calculate gradients to drive the adaptation process, and use gradient filters which are the transpose of the original filter. However, there are some slight differences in the use of

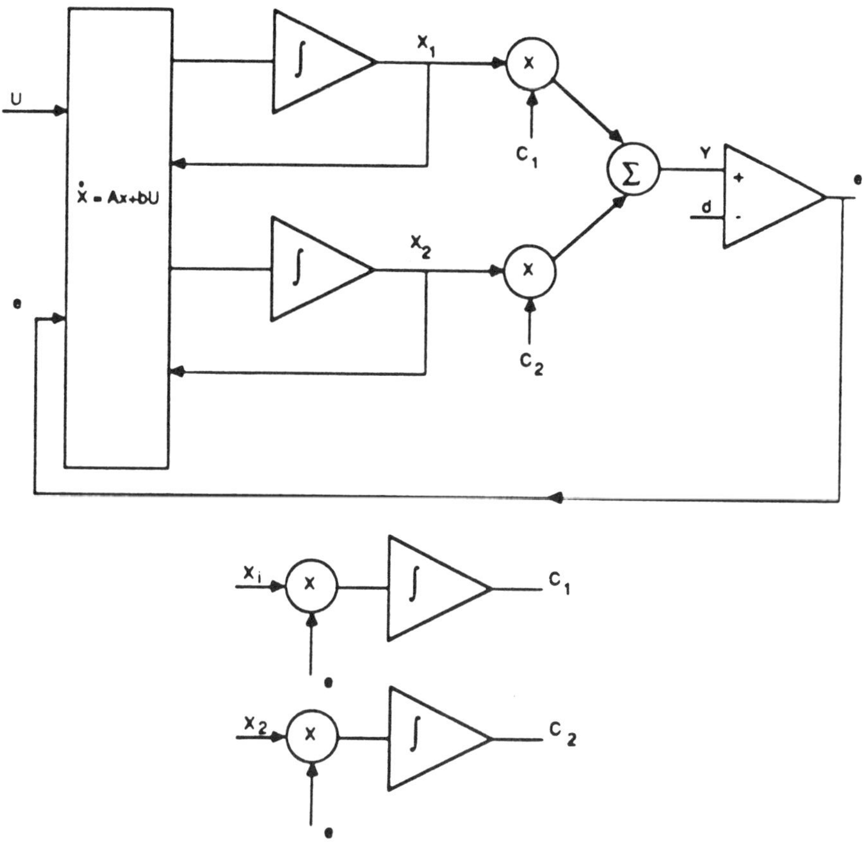

Figure 5: Analog, state space filter.

back propagation and analog gradient filters; for example, the equivalent analog circuit using back propagation uses a simpler computation of the gradient than the sensitivity approach to analog recursive filters.

In gradient adaptive filters both the neural outputs and the weights are state variables. However, the weights change much more slowly in time than the neural outputs, and these types of systems have been described in the control literature as two time scale or singular systems [20]. Since the time scales are dramatically different, the adaptation rules can take advantage of the weaker coupling between the weights and the neural outputs than the coupling among the neural outputs themselves. If the system were linear, we would say that the poles corresponding to the adaptation process are many orders of magnitude smaller than the poles that are in the frequency range of the neural inputs and output signals. The high frequency pole positions are adapted to achieve the desired signal processing, and these pole positions are varied by changing the weight integrator outputs. A similar circumstance is found in automatic tuning schemes for continuous time filters [21]. The neural circuits, however, are nonlinear systems, so instead of pole positions (time constants of complex exponentials), it is more intuitive to think in terms of generalized time constants, which are time constants of functions that are the solutions to the nonlinear differential equations of the neural system.

Neural Network Circuits

Simple continuous-time all MOS analog cells suitable for the implementation of adaptive neural circuits will be presented. An all MOS implementation of a Hopfield-like neural architecture will also be discussed. The new neural implementations take advantage of recent developments in both continuous-time MOS circuit design and advances in MOS technology.

A Continuous-Time MOS Transconductor

Here we present the simple continuous-time MOS transconductor of figure 6 as a basic parameterized analog cell [22]. It comprises four identical MOS transistors, that could be enhancement- or depletion-type, n- or p-channel, operating in the triode region. The triode region drain current of n-channel MOS device is given by

$$\begin{aligned} I_D \;=\; & \frac{W}{L}\mu C_{ox}\,\{(V_C - V_B - V_{FB} - \phi_B)(V_1 - V_2) \\ & - \frac{1}{2}[(V_1 - V_B)^2 - (V_2 - V_B)^2] \\ & - \frac{2}{3}\gamma\,[(V_1 - V_B + \phi_B)^{3/2} - (V_2 - V_B + \phi_B)^{3/2}]\Big\} \end{aligned}$$

where W and L are the channel width and length, respectively, μ is the carrier effective mobility, V_{FB} is the flat-band voltage, C_{ox} is the gate oxide capacitance per unit area, γ is the body effect, and ϕ_B is the approximate surface potential in strong inversion for zero backgate bias. The voltages V_1, V_2, V_C, and V_B are the drain, source, gate, and substrate voltages, respectively, all defined with respect to ground.

The 3/2 power terms in the above expression can be expanded in Taylor series resulting in the following compact expression for the drain current which consists of a linear term and a nonlinear term

$$I_D = \frac{1}{R}[V_1 - V_2] - [g(V_1) - g(V_2)]$$

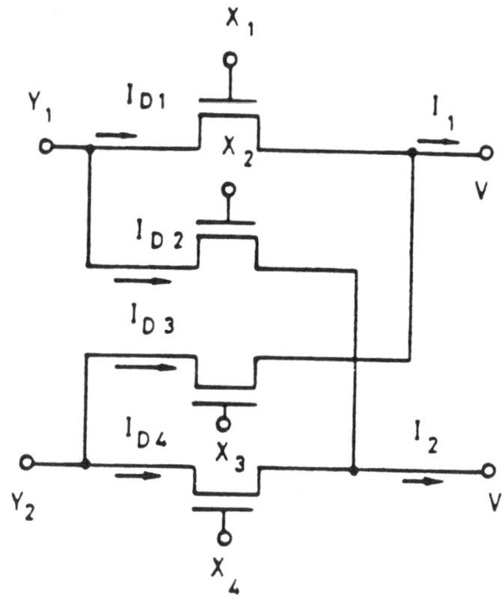

Figure 6: A versatile continuous-time MOS transconductor.

where $g(V_1)$ and $g(V_2)$ are nonlinear functions in V_1 and V_2, respectively, and are independent of the gate voltage V_C, and R is the small siganl linear resistance of the MOS transistor and is given by [23]

$$R = \frac{1}{\mu C_{ox}\frac{W}{L}(V_C - V_{TB})}$$

Now consider the MOS circuit of figure 6, driven by the voltages Y_1, Y_2, X_1, X_2, X_3, X_4, and a common mode (virtual short) voltage V. A simple analysis shows that the nonlinear components of the currents I_1 and I_2 are identical. This means complete cancellation of nonlinearities in the current difference $I_1 - I_2$ given by [22].

$$I_1 - I_2 = \mu C_{ox}\frac{W}{L}[(X_1 - X_2)Y_1 + (X_3 - X_4)Y_2 + (X_2 + X_4 - X_1 - X_3)V]$$

The nodes V are usually connected to the input terminals of a presumably ideal op-amp. The voltage V is a nonlinear function of the signals Y_1 and Y_2. Therefore, for a truly linear current difference $I_1 - I_2$, the V term must be forced to zero, resulting in the following nonlinearity cancellation condition

$$X_1 + X_3 = X_2 + X_4$$

where the general conditions for the MOS devices to operate in the triode region is given by

$$Y_i \leq \min(X_j - V_{TB})$$

where for $i = 1,\ j = 1,2$ and for $i = 2,\ j = 3,4$.

Now if we think of Y_1 and Y_2 as being AC input signals, and if we use X_1 and X_2 as DC voltages with $X_1 = X_4$ and $X_2 = X_3$, a linear transconductance G_{eq} is obtained as follows:

$$G_{eq} = \frac{i_1 - i_2}{y_1 - y_2} = \mu C_{ox} \frac{W}{L} (X_1 - X_2)$$

where

$$y_1, y_2 \leq \min[X_1 - V_{TB},\ X_2 - V_{TB}]$$

to ensure linear operation for all MOS devices. Note that a dual transconductance element can be obtained by reversing the roles of X and Y. In this case, $G_{eq} = \mu C_{ox} W/L(Y_1 - Y_2)$. It is also interesting to note that G_{eq} is independent of the threshold voltage V_{TB} and can be varied over a wide range without affecting the input signal handling capability. It also minimizes the effects of threshold voltage mismatch and substrate noise. Furthermore, it has been shown [23] that the four-transistor structure is insenstive to high-frequency parasitic capacitances.

A Simple Four-Quadrant All-MOS Vector Multiplier

An all-MOS vector multiplier is an important building block for the implementation of adaptive neural networks in analog MOS VLSI. It can easily be implemented using the transconductance element of figure 6 as a basic cell. The application of the transconductance element in the hardware implementation of a simple two-input four quadrant multiplier is illustrated in figure 7. In addition to the basic transconductance cell, the circuit contains a CMOS op-amp and two identical resistances, R [24]. The output voltage, V_0 is given by

$$V_0 = \left[\mu C_{ox} \frac{W}{L} R\right] \Delta X \cdot \Delta Y$$

where

$$\Delta X = X_1 - X_2 \quad \text{and} \quad \Delta Y = Y_1 - Y_2.$$

So, the circuit achieves four-quadrant multiplication of differential inputs as long as all transistors are operating in the triode region. An all-MOS implementation in which the resistors R are replaced by MOS transistors can be obtained by using the Double-MOSFET method [23] for nonlinearity cancellation, as illustrated in figure 8. The equivalent MOS resistance is given by

$$R = \frac{1}{\mu C_{ox} \frac{W}{L} (V_{C1} - V_{C2})}$$

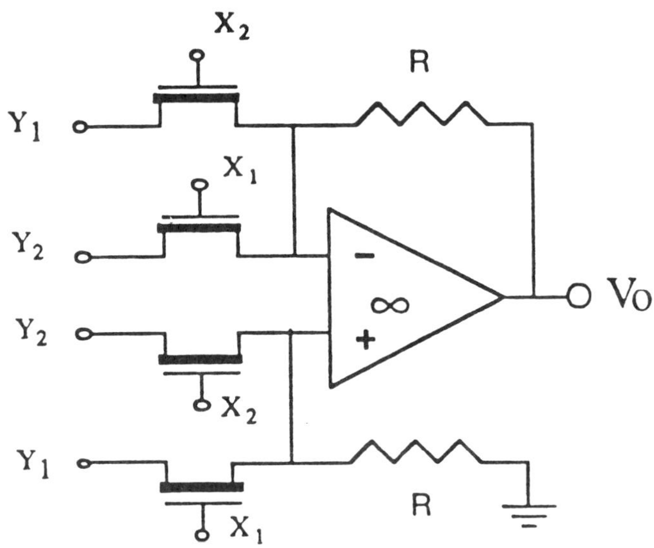

Figure 7: A simple all-MOS four-quadrant multiplier

which is designed by the proper sizing of the W/L ratio of the identical four transistors and can be controlled by the DC gate voltages V_{C1} and/or V_{C2}. The method is based on the fact that the four-transistor transconductance element can be used to simulate the current difference, $I_1 - I_2$, in a pair of resistors R as shown in figure 8. This is an interesting observation since it can be generalized to establish necessary topological requirements of a classical continuous-time circuit for its conversion to an all-MOS implementation. These requirements [23] can be stated as: (1) Resistors must occur in matched pairs, (2) voltages at one end of the pair must be the same (virtual short), and (3) voltages at the other end must be different. The Double-MOSFET method is directly applied to the simple two-input multiplier of figure 7 to obtain an all-MOS implementation [24]. The scalar product of 2 n-tuple vector inputs can easily be achieved in MOS technology by a straightforward extension of the simple two-input multiplier circuit as shown in figure 9. The output V_o of the new vector multiplier circuit is given by:

$$V_o = \frac{1}{(W/L)_0(V_{C1} - V_{C2})} \sum_{i=1}^{n} (W/L)_i \ \ \Delta X_i \cdot \Delta Y_i$$

where ΔX_i and ΔY_i are floating differential inputs given respectively by $X_{i1} - X_{i2}$ and $Y_{i1} - Y_{i2}$ and X_{ij} and Y_{ij}, $j = 1, 2$, are input voltages referred to ground. Each $\Delta X \cdot \Delta Y$ product is achieved using four identical input transistors with an aspect ratio $(W/L)_i$. The four MOS transistors M_o having an aspect ratio

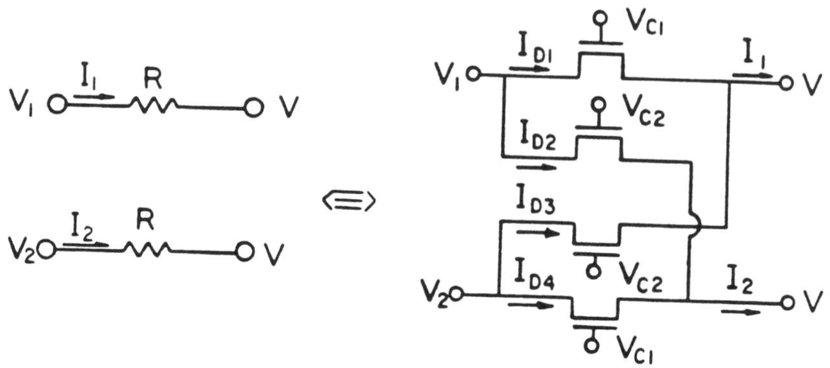

Figure 8: The Double-MOSFET method.

$(W/L)_0$ are used to replace the pair of resistors R in figure 7 according to the Double-MOSFET method. It is interesting to note that V_0 can be programmed by varying the control voltages, V_{C1} and/or V_{C2}. As we mentioned earlier the MOS implementation may use enhancement or depletion transistors, $n-$ or $p-$channel. The advantage of depletion MOS transistors, however, is the fact that they can accept positive, zero, or negative voltages at their gates as well as their drain or source terminals as evidenced by their terminal characteristics shown in figure 10. As a result, multiplication of input voltages that are referred to ground is easier with depletion transistors. If enhancement MOS devices are used the DC level of these input voltages should be shifted for proper operation [22]. The disadvantage of depletion devices is the need for an additional masking step for the implanted channel layers. The application of the MOS vector multiplier in the implementation of Hopfield-Like feedback/feedforward neural networks is discussed next.

MOS Implementation of Hopfield-Like Neural Networks

In this section we introduce a new all-MOS continuous-time implementation of the synaptic weights for Hopfield-Like feedback neural networks. The implementation is achieved via an adaptation of the MOS multipliers presented earlier where the weights are assigned as positive or negtive voltage levels. The

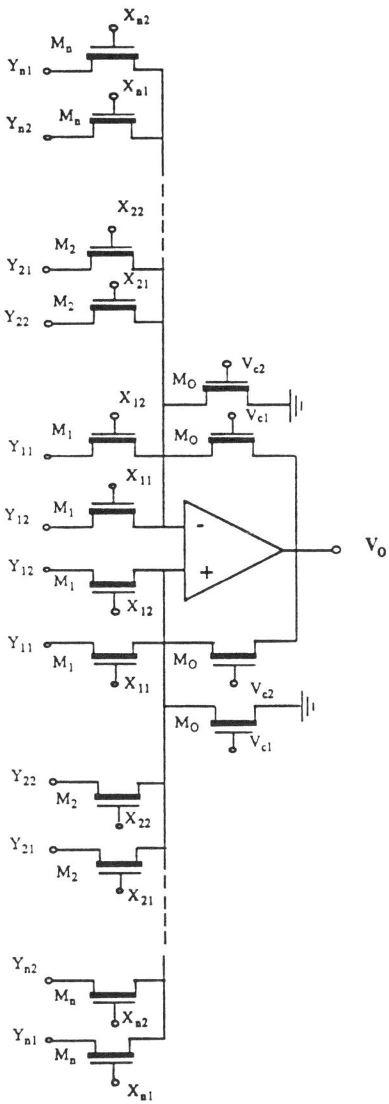

$$V_0 = \frac{1}{[(W/L)_D(V_{c1} - V_{c2})]} \sum_{i=1}^{n} (W/L)_i X_i Y_i$$

$$X_i = X_{i1} - X_{i2}, \quad Y_i = Y_{i1} - Y_{i2}$$

$$X_{ij} \text{ and } Y_{ij}, \; j = 1,2 \qquad 1 \le i \le n$$

Figure 9: A programmable Analog MOS Vector Multiplier.

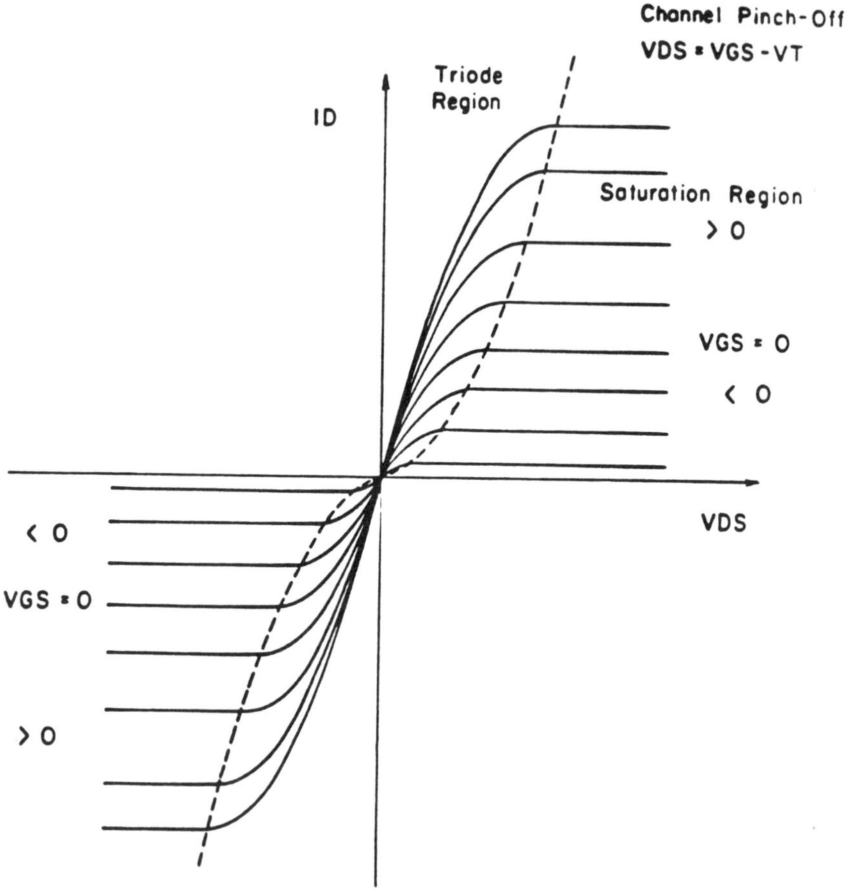

Figure 10: Depletion NMOS terminal characteristics.

neurons are realized by simple CMOS double inverters which are interconnected through the MOS vector multipliers. Each multiplier implements the scalar vector product of the vector of neuron outputs and the vector of the corresponding weights. For a network of n neurons, there are n such scalar products. Each scalar product is achieved using only one operational amplifier and $4(n+1)$ MOS transistors for 2 n-tuple vector inputs resulting in an economic and attractive analog MOS VLSI implementation. A neuron and its associated vector multiplier are illustrated in figure 11. Using depletion transistors, gates of MOS transistors can be connected to ground resulting in a special case of the vector multiplier presented earlier which allows the multiplication of voltages that are referred to ground. Positive or negative grounded voltage levels can be assigned to the synaptic weights, Y_i. The outputs of n neurons X_i are fedback as inputs to the ith multiplier $(1 \leq i \leq n)$. The output of the ith multiplier in turn is fed into the input of the ith double inverter (neuron i). The output of the two-input multiplier shown in the dotted subsection in figure 11 is proportional to $X_1 Y_1$ where X_1, the output of neuron 1, and Y_1, its associated weight, are voltages referred to ground. The overall output of the vector multiplier, V_0 is given by

$$V_0 = \frac{2}{(W/L)_0(V_{C1} - V_{C2})} \sum_{i=1}^{n} (W/L)_i X_i Y_i$$

All MOS transistors must be operating in the triode region. Hence at the multiplier's input

$$|Y_i| \leq |V_{TB}| \quad \text{for positive } X_i$$

and

$$-X_i + |Y_i| \leq |V_{TB}| \quad \text{for negative } X_i$$

It is interesting to note that X_i, when it is positive, is restricted only by the maximum allowable gate voltage specified in the process electrical design rules. At the multiplier's output, V_0 should satisfy

$$V_0 \leq \min[V_{C1} - V_{TB},\ V_{C2} - V_{TB}].$$

The input-output compatibility of the overall MOS implementation is of particular interest since the relatively high output impedance node of the double inverter is connected to the almost infinite input impedance of the MOSFET gates with almost no restrictions on the fan-in/fan-out capability. Another function of the double inverter is to limit the dynamic range of the inputs on the neural output using the inverters sigmoidal nonlinearity. However, the double inverter can be removed and the saturating nonlinearity of the multiplier itself can be utilized. The onset of saturation can be set using the multiplier resistor pair. The maximum value of n is limited when we use the four-transistors to replace the multiplier resistor pair. It is determined by the maximum currents these transistors can carry. This is a situation that can be created when the

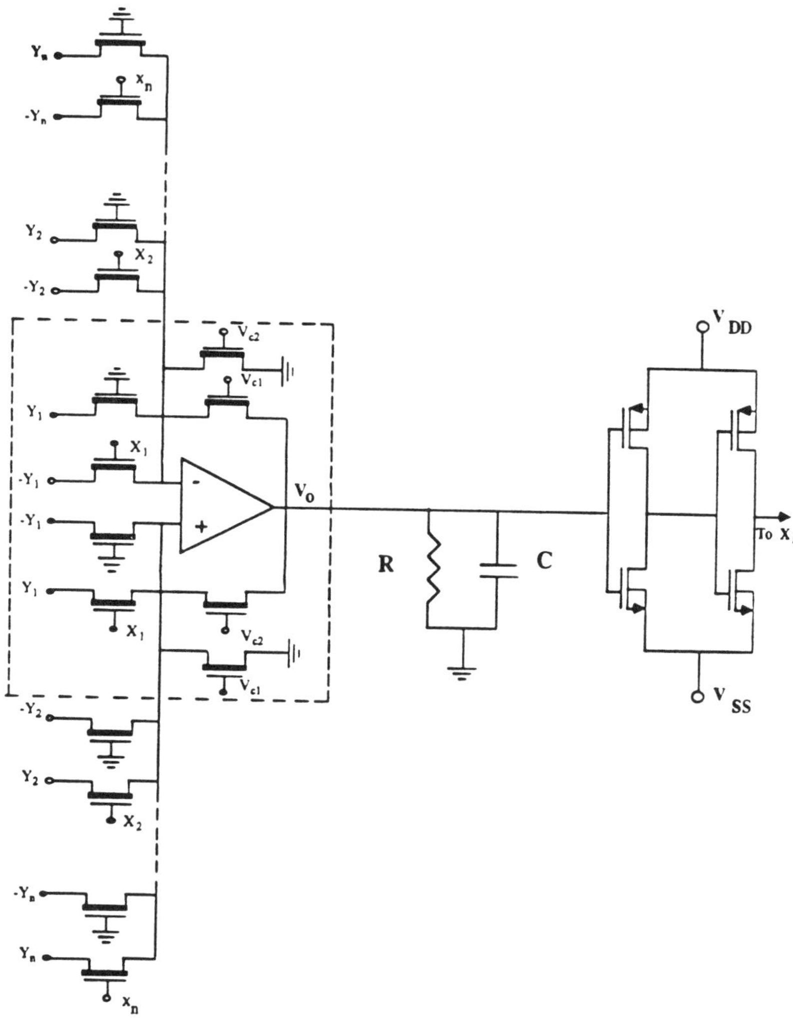

Figure 11: A MOS circuit for the VLSI Implementation of Hopfield-like nerual networks.

polarities of X_i and Y_i are such that all input transistor currents are coming in (or out of) the op-amp summing input nodes. The sizing of the feedback transistors in multiplier can be analyzed using the previous equations to set the dynamic range of the linear region. For an overall implementation that uses only enhancement transistors [26], the output signals of the neurons must be level shifted properly before they are fed back to the multiplier inputs. This can be easily achieved using a simple MOS DC level shifter [26].

It is also interesting to note that in the well-known Hopfield circuit [14], complement outputs of the neurons are used in the feedback since negative passive resistors are impossible to implement. The MOS implementation described here has a single output for each neuron. This can reduce the problems of VLSI routing and interconnects. Computer simulations [25,26] of the MOS implementations verified the theoretical development and exhibited the robustness properties of neural networks. The ideas and concepts presented can be applied equally well to implement feedforward neural architectures.

Programmable Threshold Voltage Device Circuits

In order to maximize the size of the neural system which can be put on silicon chips, it is desirable to combine as much of the neural functions as possible into a given circuit module. One method of combining the integration, or storage, of a signal with a multiplying function is to use a single transistor with an alterable threshold voltage. The multiplicative computation involves the gate and drain voltages to produce a transistor current.

A alternate approach to developing a minimum size weight integrator is to use a transistor amplifier with a Miller capacitor. However, in some adaptive systems, adaptation rates on the order of seconds or minutes are needed; and in the case of associative memories, we would like to extend the weight integrator to become an analog memory unit. An analog memory element (AME) is useful if the neural circuit is to be used for pattern storage or to compensate signals from deleterious component offsets resulting from the IC fabrication technology.

Therefore, in order to develop long term analog storage, the threshold voltage of a MOS transistor can be varied by use of charge tunnelling and trapping effects in either floating gate [27], or MNOS devices [28]. Threshold shifts can be obtained on standard MOSIS technology parts, either by using UV light [29], or by using liquid nitrogen temperatures [30]. In a double poly process, both drain avalanche and control gate induced tunnelling can be utilized.

Shown in figure 12 is a four neuron, 16 weight synapse circuit in which threshold voltages are to be altered [31]. The double poly process offered by MOSIS is fabricating the die. In order to use only electrical programming, both gate induced and drain avalanche induced charge tunnelling are used. Each weight is put in its own well which can be independently biased to promote avalanching of its drain but not other drains on the same word line. Gate induced tunnelling allows for lowering of drain voltages to circumvent drain-substrate breakdown effects.

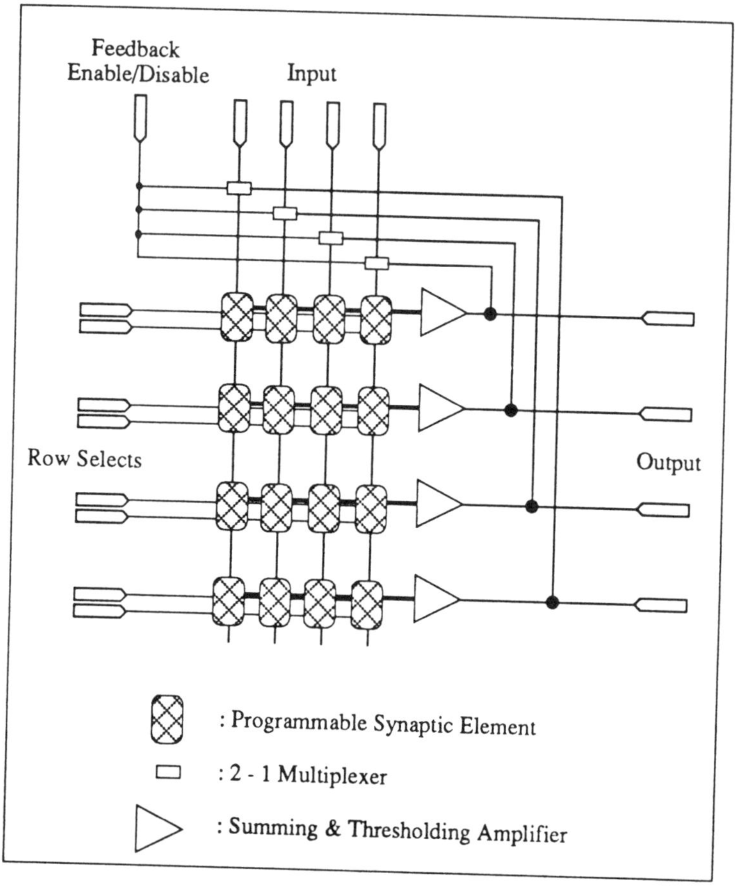

Figure 12: Floating gate neural network.

In order to increment and decrement the weight values, a pair of programmable transistors are used to develop a current difference, as shown in figure 13. The current difference from several sets of synapses is sensed by a set of current mirror loads, as shown in figure 14, where 2 neurons and four synapses are shown. The synapses can be operated in the subthreshold current region by adjusting the current mirror bias, yet the neural output is taken through a strong inversion transistor amplifier with a stronger current drive. A layout of the chip is shown in figure 15. The weight update circuitry is missing from

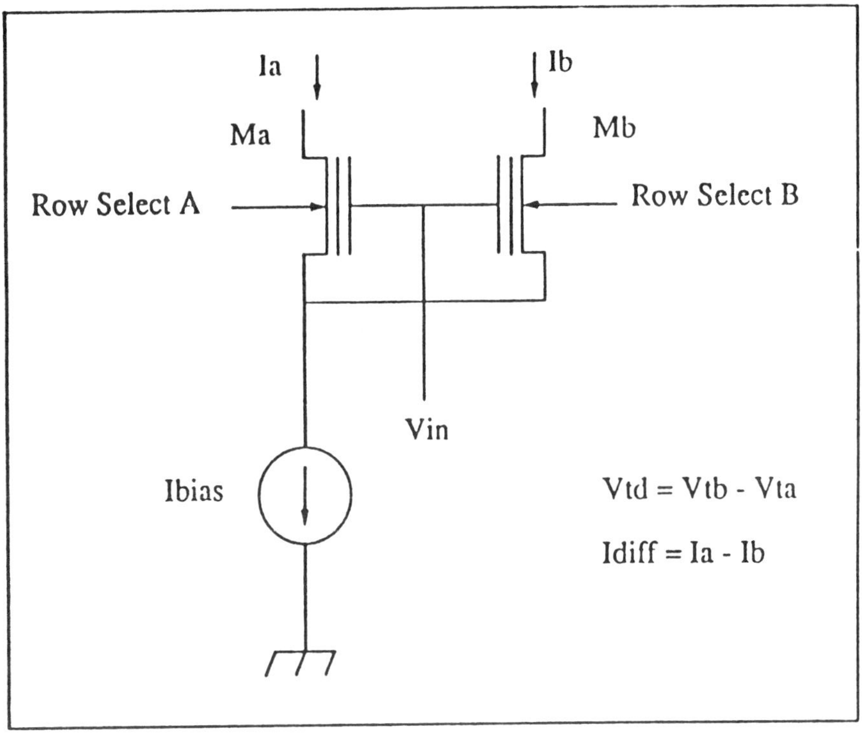

Figure 13: Pair of programmable transistors to increase or decrease weight.

the left side of this version of the chip, since the first objective is to test the electrical programming range of the floating gate transistor synapses on the right.

By sizing transistors appropriately, the neural output can be made to be a very sensitive function of the current differences, as shown in figure 16. Therefore, even small threshold voltage shifts can be utilized. The threshold voltage shift range and the sizing of the load and output circuits are the two major factors in determining the dynamic range of the weight circuits.

Conclusions

We have discussed important issues in analog VLSI that are relevant to the hardware implementation of microelectronic neural networks in MOS technology. Similarities between the neural paradigm of computation and the design of analog, adaptive, state-space filters have been outlined. The circuit techniques

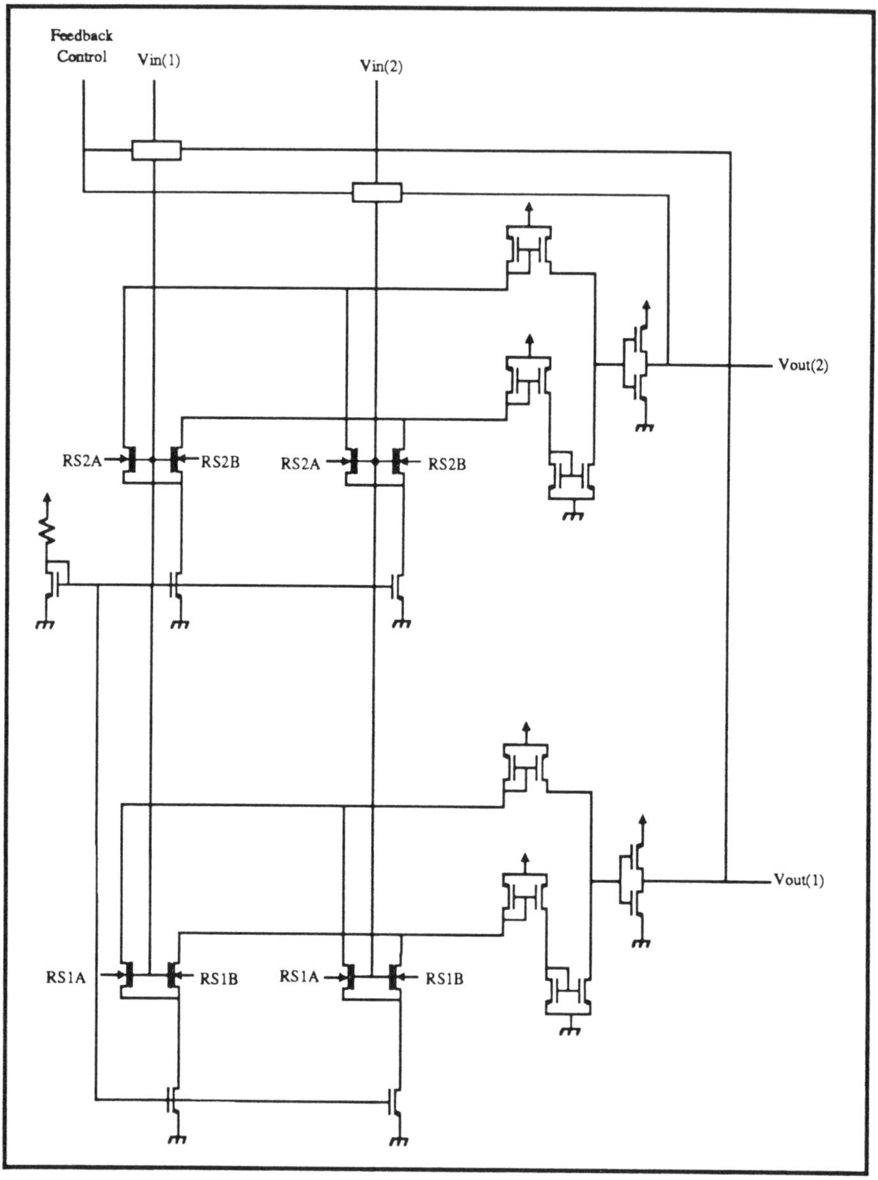

Figure 14: Two neuron circuit with four weight synapses.

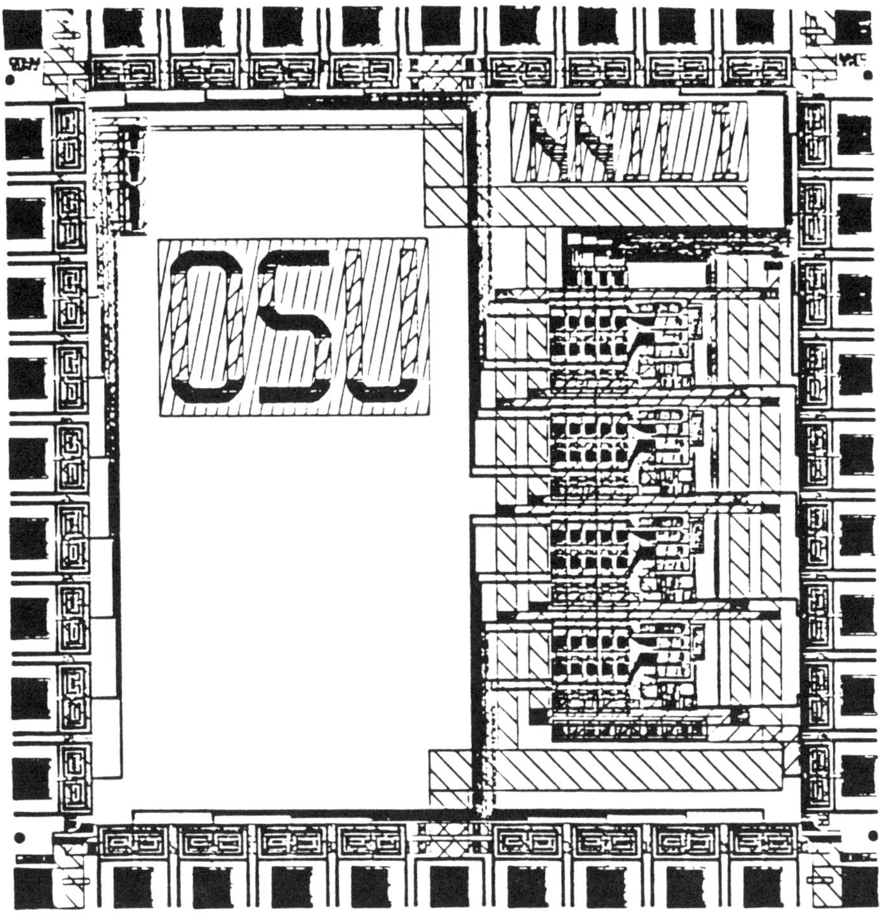

Figure 15: Layout of 4 neuron chip.

are based on simple and programmable analog parameterized cells with features of reconfigurability and input/output compatability. Analog circuit modules, such as integrators, summers, multipliers, etc., have been configured in a neural network architecture to build the equivalent of state-space signal processors. We presented a methodology for comparing various forms of adaptation based on driving weight integrators with circuitry which computes estimates of a gradient of a criterion or energy function. The methods for estimating the gradients are similar to the techniques which have been developed for traditional signal processing. The circuit implementations take advantage of continuous-time MOS circuit design concepts as well as advanced CMOS technologies.

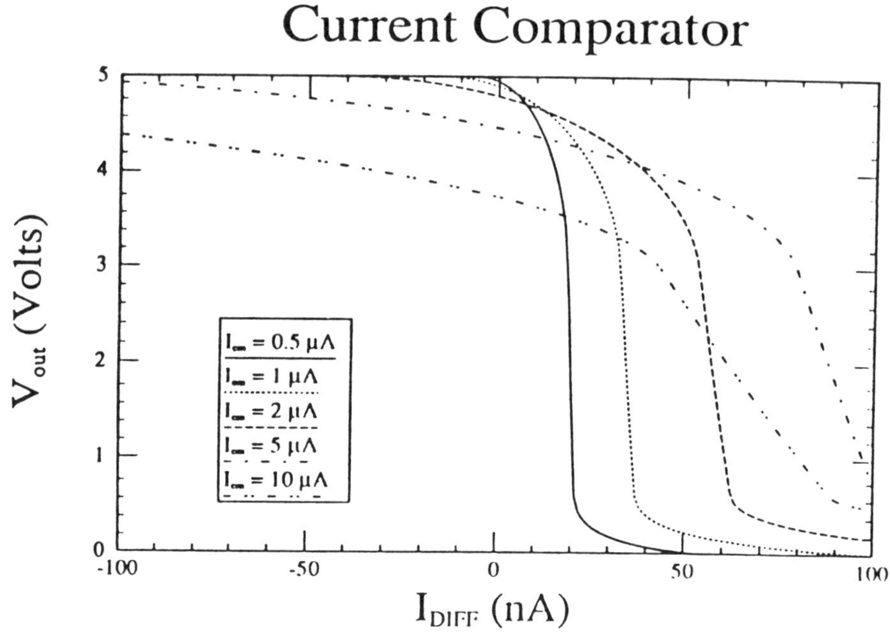

Figure 16: Neural output as a function of differential and common mode currents.

Acknowledgements

We would like to thank our many colleagues, with special thanks to Profs. P. Allen and F. Salam. The following students have done much of the work: Ken Adkins, Zaka Bhatti, Tom Borgstrom, Scott Dupuie, Rich Kaul, and Nabil Khachab. The encouragement and support of this research by NSF, the Semiconductor Research Corp., AT&T, the NASA Lewis Research Center, and the Electrical Engineering Dept. at Ohio State is much appreciated.

References

[1] C. Mead, *Analog VLSI and Neural Systems*, Reading, Mass., Addison-Wesley, 1989.

[2] M. Ismail and J. Franka, *Introduction to Analog VLSI Design Automation*, Kluwer Academic Publishers, Boston, 1989.

[3] M.R. Haskard and I.C. May, *Analog VLSI Design: NMOS and CMOS*, Prentice-Hall, New York, 1988.

[4] N. El-Leithy and R.W. Newcomb, Special Issue on Neural Networks. *IEEE Transactions on Circuit and Systems*, May 1989. Also, S. Bibyk and M. Ismail, Analog Signal Processing for Neural Microelectronics, Special Session, *Proc. IEEE ISCAS*, May. 1989.

[5] E. Habekotte et al. "State-of-the-Art in the Analog CMOS Circuit Design" *Proc. IEEE*, Vol. 75, pp. 816-828, June 1987.

[6] Y. Tsividis, "Analog MOS Integrated Circuits: Certain New Ideas, Trends, and Obstacles", *IEEE J. Solid-State Circuits*, Vol. SC-22, pp. 317-321, June 1987.

[7] M. Ismail, "Continuous-time Analog Design for MOS VLSI" State-of-the-Art Review invited paper, *Proc. of the 30th Midwest Symp. on Circuits and Systems*, pp. 707-711, Elsevier Science Publishing Co., 1987.

[8] P.R. Gray, B. Wooley, and R.W. Broderson, *Analog MOS Integrated Circuits* IEEE Press book, New York, 1989.

[9] P.E. Allen, "CAD for Analog VLSI", IEEE CAS Distinguished Lecturer Program, April 24, 1989.

[10] M. Ismail, "Reconfigurability, Versatility and Modularity in analog IC Design," presented at the Semiconductor Research Corporation (SRC) Workshop on Analog Design Automation, December 2nd, 1988.

[11] J.L. Hilbert, SRC Private Communication, December, 1988.

[12] L.R. Carley and R.A. Rutenbar. "How to Automate Analog IC Designs," *IEEE Spectrum*, pp. 26-30, August 1988.

[13] H.Y. Koh, C.H. Sequin, and P.R. Gray, "Auto Synthesis of Operational Amplifiers Based on Analytic Circuit Modes", *Proc. IEEE ICCAD*, pp. 502-505, November 1987.

[14] J.J. Hopfield, "Neurons with graded response have collective computational properties like those of two state neurons," *Proc. Natl. Acad. Sci.*,USA, vol. 81, pp. 3088-3092, May 1984.

[15] B. Widrow and S. Stearns, *Adaptive Signal Processing*, Englewood Cliffs, NJ, Prentice Hall, 1985.

[16] T. Kohonen, *Self-Organization and Associative Memory*, Second Edition, 1987.

[17] S. Bibyk and K. Adkins, "Neural Nets and Emergent Adaptive Signal Processing," *Proc. of IEEE Int. Symp. Circuits and Systems*, May 1989, pp. 1203-1206.

[18] D.E. Rummelhart, J.L. McClelland, and the PDP Research Group, *Parallel Distributed Processing*, The MIT Press, vol. 1, chp. 8, 1986.

[19] D. Johns, W. Snelgrove, and A. Sedra, "Continuous-Time Analog Adaptive Recursive Filters," *Proc. of IEEE Int. Symp. Circuits and Systems*, May 1989, pp. 667-670.

[20] V. Saksena, J. O'Reilly, and P. Kokotovic, "Singular Perturbations and Time-scale Methods in Control Theory: Survey 1976-1983," *Automatica*, vol. 20, pp. 273-293, May 1984.

[21] T.L. Brooks and P.M. VanPeteghem, "Simultaneous Tuning and Signal Processing in Integrated Continuous Time Filters: The Correlated Tuning Loop," *Proc. of IEEE Int. Symp. Circuits and Systems*, May 1989, pp. 651-654.

[22] M. Ismail, "Four Transistor Continuous-Time MOS Transconductor," *Electronics Letters*, Vol. 23, No. 20, pp. 1099-1100, September 1987.

[23] M. Ismail, S. Smith and R. Beale, "A New MOSFET-C Universal Filter Structure for VLSI," *IEEE J. Solid-State Circuits*, Vol. 23, pp. 183-194, February 1988.

[24] N. Khachab and M. Ismail, "Novel Continuous-Time All-MOS Four-Quadrant Multipliers", *Proc IEEE ISCAS*, pp. 762-765, May 1987.

[25] F. Salam, N. Khachab, M. Ismail, and Y. Wang, "An Analog MOS Implementation of the Synaptic Weights for Feedback Neural Nets," *Proc. IEEE ISCAS*, pp. 1223-1225, May 1989.

[26] N. Khachab and M. Ismail, "An Analog MOS VLSI Implementation of Hopfield-like Neural Networks," to appear.

[27] R. Shimabukuro, I. Lagnado, and P. Shoemaker, "A Dual Polarity Nonvolatile Analog Memory for Use in Adaptive Neural Networks," Silicon Nitride and Silicon Dioxide Thin Insulating Films, ed. S. Bibyk et al., *Proc. of the Electrochemical Soc.*, vol 89-7, pp 157-165.

[28] J. Sage, and R. Withers, "Analog Nonvolatile Memory for Neural Network Implementations," Silicon Nitride and Silicon Dioxide Thin Insulating Films, ed. S. Bibyk et al., *Proc. of the Electrochemical Soc.*, vol 89-7, pp 157-165.

[29] L. Glasser, "A UV Write-Enabled PROM," *1985 Chapel Hill Conference on VLSI*, pp. 61-66.

[30] S. Bibyk, H. Wang, P. Borton, "Analyzing Hot-Carrier Effects on Cold CMOS Devices," *IEEE Trans. Elec. Dev.*, vol. ED-34, pp. 83-88, Jan. 1987.

[31] T. Borgstrom and S. Bibyk, "A Neural Network Circuit Utilizing Programmable Threshold Voltage Devices," *Proc. of IEEE Int. Symp. Circuits and Systems*, May 1989, pp. 1227-1230.

6

DESIGN AND FABRICATION OF VLSI COMPONENTS FOR A GENERAL PURPOSE ANALOG NEURAL COMPUTER

PAUL MUELLER, JAN VAN DER SPIEGEL, DAVID BLACKMAN, TIMOTHY CHIU, THOMAS CLARE, CHRISTOPHER DONHAM, TZU PU HSIEH, MARC LOINAZ

Departments of Biochemistry and Biophysics and Electrical Engineering

University of Pennsylvania
Philadelphia, Pa.

SUMMARY

The design of components for a programmable analog neural computer and simulator is described. The machine can be **scaled to any size** and is composed of three types of interconnected **modules**, each containing on a VLSI chip arrays of **Neurons**, modifiable **Synapses** and **Routing Switches**. It runs entirely in analog mode but the connection architecture, synaptic gains and time constants as well as neuron parameters are set digitally from a digital host computer. Each neuron has a limited number of inputs and can be connected to any but not all other neurons.

The neuron circuit consists of a rectified summing amplifier, comparator and output driver. Inputs to the neurons are currents, outputs are analog voltages. The following neuron parameters can be adjusted through digital control: threshold (bias), minimum output at threshold and linearity of the transfer function. For the computation of synaptic weights by the host computer on the basis of learning algorithms, time segments of the neuron outputs are multiplexed, converted to digital form and stored in memory.

The synaptic weights are implemented by current mirrors that scale the neuron outputs after they have been converted linearly from a voltage to a current. The weights are set by serial input from the host computer and are stored at each synapse. Dynamic range of the weights extends from 0 to 10 with 5 bit logarithmic resolution; a sixth bit determines the sign. Synaptic time constants are programmed at the inputs to the synapse line.

The routing switches connect vertical and horizontal lines of a cross point array and also can cut these lines. Each switch cell is implemented as a transmission gate connected to one bit of memory. The switches are set by serial input from the host computer.

The machine is intended for real-world, real-time computations such as vision, acoustics or robotics and the development of special purpose neural nets. Even at moderate size of 10^3 to 10^5 neurons the computational speed is expected to exceed by orders of magnitude that of any current digital computer.

INTRODUCTION

The computation of real world phenomena in real time requires computational power that exceeds by many orders of magnitude the capabilities of sequential digital machines.

Biological brains are able to solve tasks such as seeing or hearing because they operate in analog mode which allows simultaneous summing of many inputs from interconnected units and permits large scale parallel processing without the need for iterative procedures.

Extrapolation from simulations of simple neural circuits indicate that a sequential digital machine would have to operate at speeds of more than 10^{18} floating point operations per second in order to match the performance limit of the human brain.

The advantages of neural computation are now widely recognised and electronic implementation of neural systems based on analog circuits of neurons and synapses is currently being pursued in a number of laboratories[1-14] where several special purpose systems have been fabricated in VLSI[1,8,12-17] or macro components[18].

Unfortunately, most of the connection architectures and computational strategies implemented by biology are not yet known and it seems desirable to have available a general purpose machine in which the connections as well as the component parameters - such as neuron thresholds and transfer functions, synaptic gains and time constants can be programmed either externally or by the machine itself.

This paper describes the design of a general purpose analog neural computer and presents performance data of VLSI modules for the machine. A preliminary report has been published elsewhere.[7]

OVERVIEW OF THE NEURAL COMPUTER

Before discussing the details of component design, we shall give a general overview of the computer.

The machine architecture, shown in Fig. 1, is loosely based on the cerebral cortex in the sense that there are separate neurons, axons and synapses and that each neuron can receive only a limited number of inputs. However, in contrast to the biological system, the connections can be modified by external control permitting exploration of different architectures in addition to adjustment of neuron parameters and synaptic weights.[7] The design has evolved from our previous experience with manually programmable neuron nets for the analysis of acoustical patterns[18,19].

The machine contains large numbers of the following separate elements: **neurons, synapses, and routing switches.** Arrays of these elements are fabricated on VLSI chips that are mounted on planer chip carriers each of which forms a separate **module**. The modules are connected directly to neighboring modules on a circuit board. Neuron arrays are arranged in rows and columns and are surrounded by synaptic and routing switch arrays. The switches select the connections between neurons. The direction of data flow is shown in Fig. 2.

The computer runs entirely in analog mode. However, connection architectures, synaptic gains and neuron parameters such as thresholds and time constants are set by a digital host computer either directly from the keyboard or from stored programs.

For the implementation of learning algorithms, time segments of the outputs from all neurons are multiplexed, digitized and stored in memory of the host computer. The stored outputs are used to compute the adjusted synaptic weights that are then set by the computer. The multiplexing operation is independent of, and does not interfere with the actual analog computations.

The modular design allows expansion to any degree and at moderate to large size, i.e. 10^3 to 10^5 neurons, operational speed would exceed by 3 to 6 orders of magnitude that of any currently available digital computer.

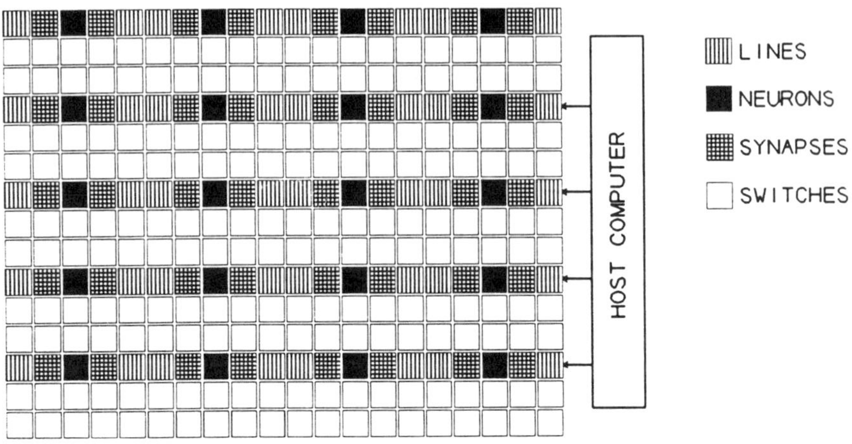

Fig. 1. Layout and general architecture. The machine is composed of different modules shown here as squares. Each module contains on a VLSI chip an array of components (neurons, synapses or switches) and their control circuits. A prototype design would contain 64 neuron modules for a total of 1024 neurons each having 64 synapses. The symmetry of the connections between modules allows adjustment of the ratio between different modules and unlimited addition of modules.

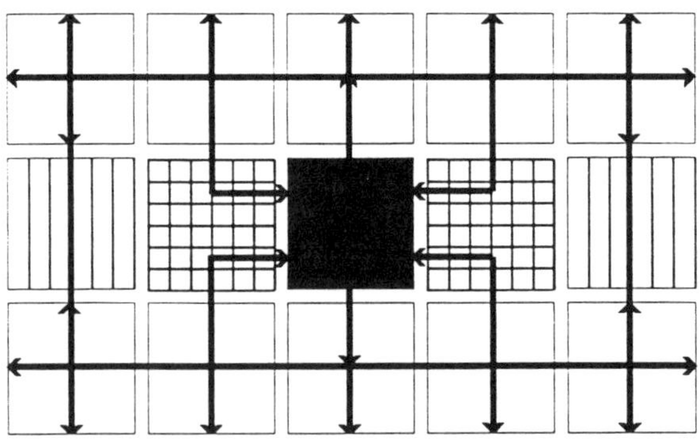

Fig. 2. The direction of data flow through the modules. Outputs from each neuron leave north and south and are routed through the switch modules east and west and into the synapse modules from north and south. They can also bypass the synapse modules north and south via lines. Input to the neurons through the synapses is from east and west. Power, digital control lines and multiplexed neuron outputs run east and west. The multiplexed outputs to the host computer are not shown.

THE COMPONENT MODULES

In the following sections we discuss the design of the individual VLSI modules. All chips are designed for processes available through MOSIS using the Berkeley VLSI tools.

Several versions of small prototypes have been fabricated in 3u and 2u CMOS, and test results are presented. The chip control circuits and operation are also discussed.

The Neuron Module

Each neuron chip contains 16 neurons, an analog multiplexer and control logic (see Fig. 3).

Input-output relations of the neurons are idealized versions of a typical biological neuron. The circuit and the neuron transfer function, which are based on an earlier design using discrete components,[18] are shown in Fig. 5-8. The circuit consists of a rectified summing amplifier, comparator and output driver. Each unit has an externally adjustable threshold (bias), an adjustable minimum output value at threshold, E_x, a linear transfer function above threshold and a maximum output (see Figs.4 and 9). Output time constants are selected on the switch chips (see below). The output has only one sign; positive or negative input polarity is selected at each synapse.

The output buffers were designed to drive a resistive load of <1 KOhm and a capacitive load of ~ 100 pF. The actual resistive load will be much less because the synaptic inputs which the neuron drives are FET gates and the leak resistance to ground of the routing switches is > 10^{12} Ohm. As tested, the gain-bandwidth product of the output buffer was 900 KHz, the slew rate $5x10^6$ V/s, settling time 14 us and the phase margin 35 degrees. These test results agreed well with the SPICE simulations.

Inputs to each neuron come from synapse chips east and west (SIR, SIL), outputs (NO) go to switch chips north and south. Each neuron has a separate input that sets the minimum output at threshold (E_x) which is common for all neurons on the chip and selected through a separate synapse line. The threshold is set from one of the synapses connected to a fixed voltage.

In learning mode the synaptic gains and neuron parameters are computed and set from the control computer on the basis of the neuron outputs. For this to be possible, an analog multiplexer provides neuron output to a common line, OM, which connects to a fast A/D converter and stores selected time segments of all neuron outputs in memory. The switches of the multiplexer are controlled by a shift register clocked via two phase clocks and master clock, CK. By passing control pulses (ORO and ORI) from chip to chip all neurons are read in sequence every 2 ms. For discussion of the A/D conversion see below.

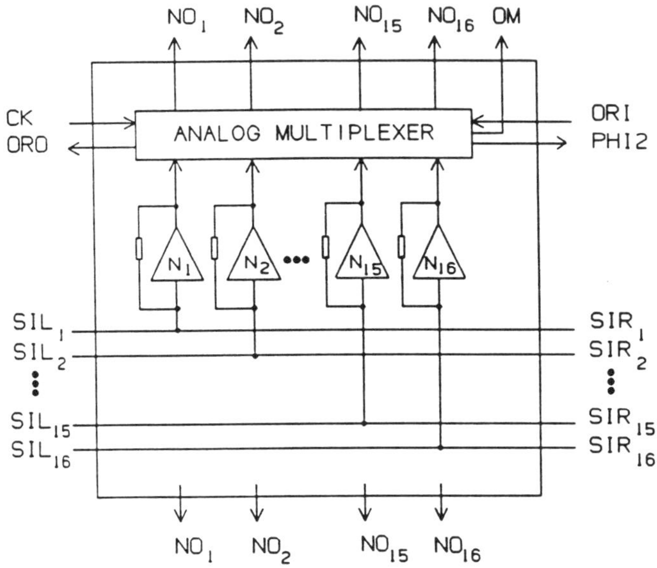

Fig. 3. Block diagram of the neuron chip containing 16 neurons.

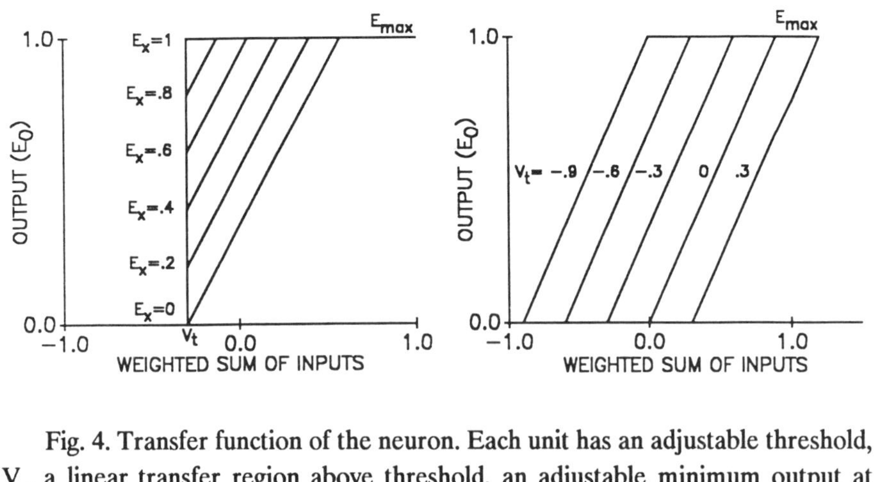

Fig. 4. Transfer function of the neuron. Each unit has an adjustable threshold, V_t, a linear transfer region above threshold, an adjustable minimum output at threshold E_x and a maximum output, E_{max}. The adjustment of $\mathbf{E_x}$ and of $\mathbf{V_t}$ are shown in **A** and **B**.

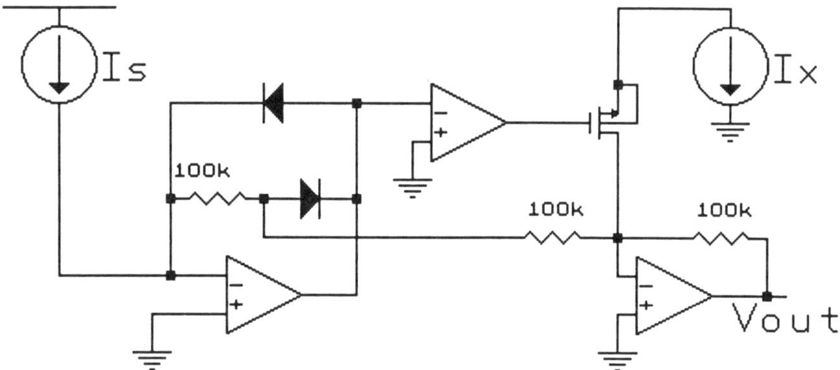

Fig. 5. Neuron circuit. The extra output at threshold, E_X, is controlled by I_X. Two different versions of this circuit have been fabricated. The first version was designed to drive resistive loads and had two outputs of opposite sign. The later version had only one output designed for capacitive loads. The layouts of both versions are shown in fig. 7.

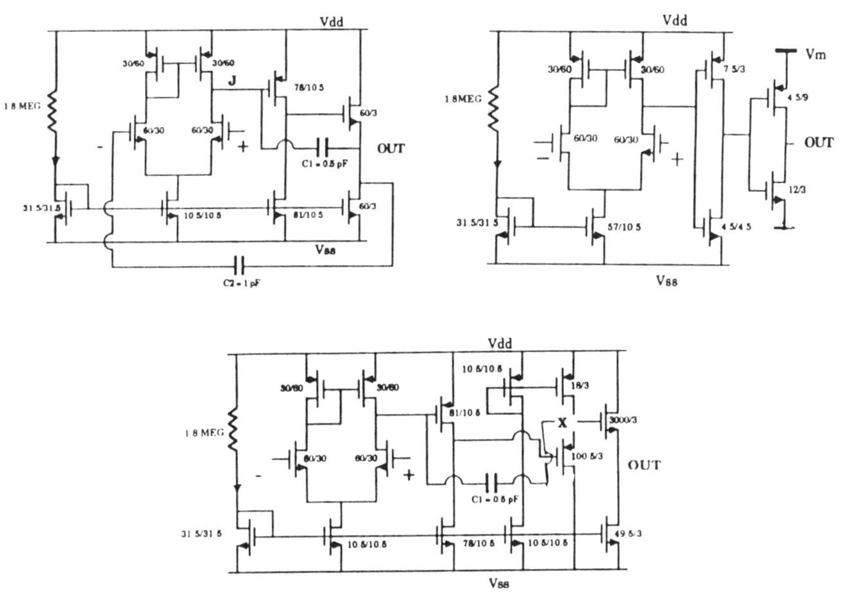

Fig. 6. Operational amplifiers of the neuron circuit.

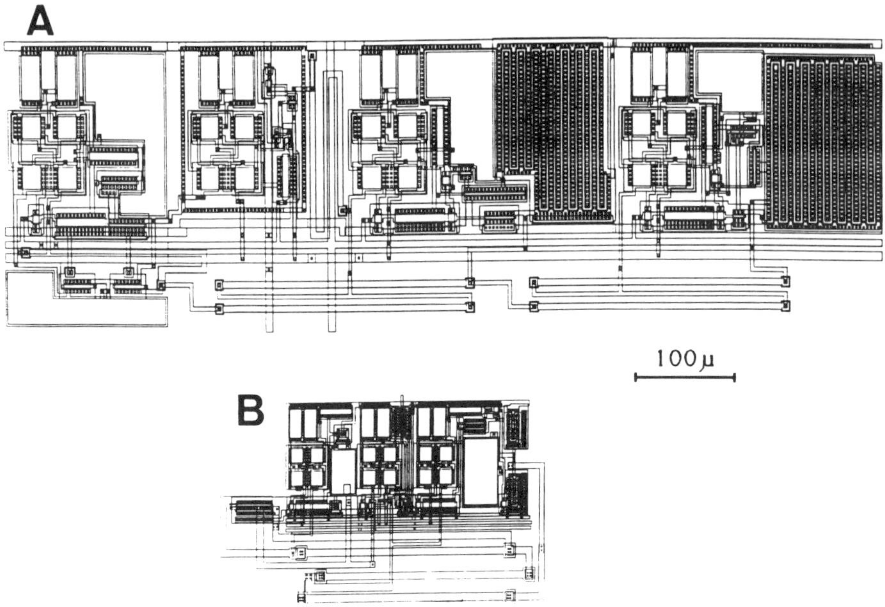

Fig. 7. Layout of two versions of the neuron circuit. The older version is shown in **A**. Notice the big output drivers designed for large resistive loads. The later version, **B**, has only one output. It was fabricated in 2 u CMOS.

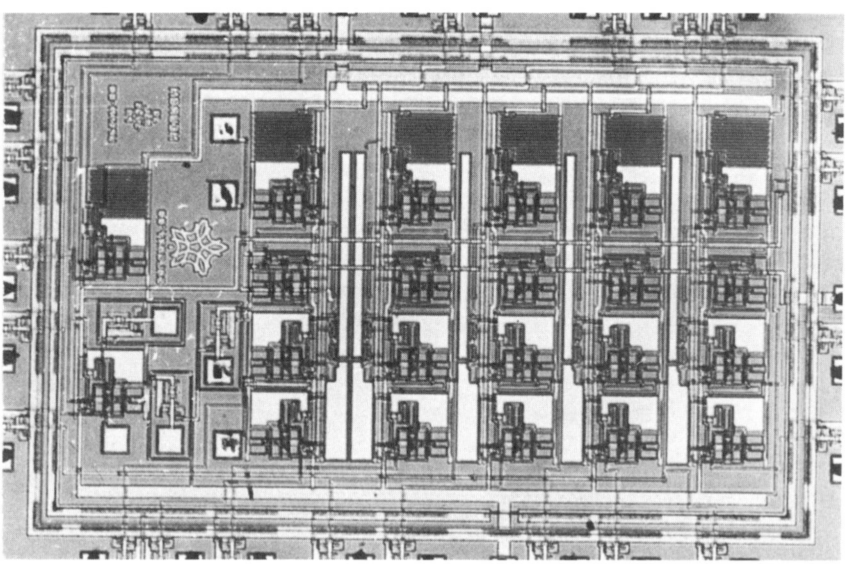

Fig. 8. Photograph of a test chip containing 5 neurons. This is the older version of the circuit.

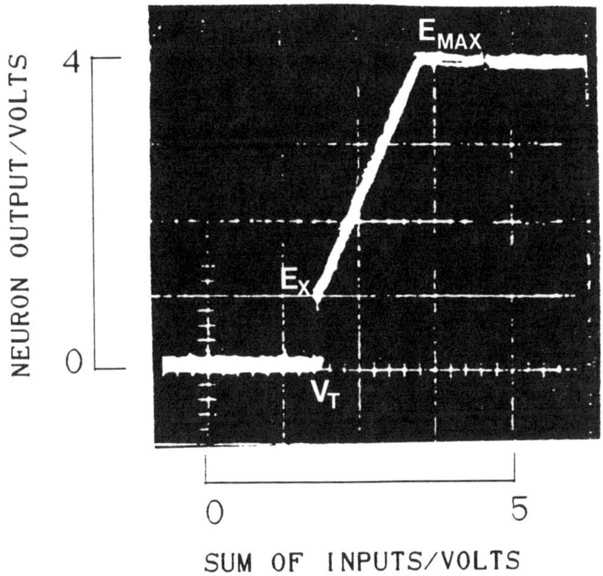

Fig.9 Transfer characteristic obtained from a neuron on the chip shown in Fig. 8. The threshold, V_t was set here to 1.5 V,and the adjustable minimum output at threshold E_x to 1 V .

TABLE 1

NEURON SPECIFICATIONS

Process	2u CMOS
Operation	Current Summing
Outputs	0 to 4 V
Transfer Function	Linear[1] or sigmoid [2]
Output Impedance	< 1 KOhm
Gain- Bandwidth Product	900 KHz
Settling time	14 us
Threshold Adjustment	Logarithmic, + or - , 5 bit resolution
Adjustment of Ex	0 to 4 V, 5 bits

1) Currently implemented , 2) Planned.

Discussion Of The Neuron Properties

Several aspects of the neuron design require comment.

1. Threshold adjustment

The threshold of each neuron is individually adjustable from the synapse chip via an extra synapse that is connected to a fixed input voltage. In this way the threshold can be biased in either direction. A neuron with a negative threshold bias produces output in the absence of inputs from other neurons. This feature is often observed in biological neurons and is also important for certain learning algorithms such as backpropagation.

2. The minimum output at threshold

Each neuron has an adjustable minimum output at threshold, $\mathbf{E_x}$ that can be set to any value between 0 and $\mathbf{E}_{max}$ by application of a current to pin $\mathbf{E_x}$ at the comparator circuit (see Fig. 5). This adjustment is the same for all neurons on one chip. The current is derived from a synapse circuit on the synapse chip which generates a selected current.

The adjustable minimum output at threshold is an important feature which enables the neuron to perform logic as well as arithmetic operations by the same unit. In the limit the neuron can function either as a boolean switch when $\mathbf{E}_x = \mathbf{E}_{max}$, or as a summing amplifier when $\mathbf{E}_x$ is set to 0. Intermediate settings permit combined logic and arithmetic operations (transparent computation) by the same unit. This feature is also found in biological neurons which in many cases begin firing at a relatively high rate when the sum of inputs reaches the threshold.

3. The input - output transfer function.

As currently implemented, the I/O transfer function is linear between $\mathbf{E}_x$ and $\mathbf{E}_{max}$ (see Fig.4). This conforms roughly to the relation between average membrane depolarization and firing rate of a typical biological neuron tested in isolation. In most situations the linearity of the I/O function is not critical but our experience with a network for acoustical pattern recognition[18] showed that the linearity contributes to ease of programming and stable operation especially in the time domain.

There are reasons, however, for looking at alternative transfer functions. Specifically we shall consider designing neurons with a sigmoid transfer function. This transfer function is widely used in the simulation of learning algorithms, such as backpropagation,[20] where it has proven especially effective. A sigmoidal transfer function can be obtained by adding one or more non- linear devices, e.g. transistors in parallel with the feedback resistor of the summing opamp.

In order to investigate this point in more detail we have carried out simulations of simple learning tasks using a backpropagation algorithm. In the case of learning an XOR function by a 5 neuron network the sigmoid transfer function performed considerably better for identical initial conditions (less iterations and fewer failures to converge). However, linear transfer functions with $E_x = 0$ or with $E_x > 0$ gave acceptable results provided that the derivative of the function was assumed to be that of a sigmoidal. Further tests on this point are planned.

In addition to its utility for gradient descent learning, a sigmoidal transfer function would also enable the neurons to perform multiplication and division of different inputs by biasing the operating region into the exponential or logarithmic portions of the sigmoid.

4. Are spiking neurons necessary?

The neurons shown in Fig. 5 do not generate action potentials but transmit instead their output voltages as continuous variables. Nerve fibers have extremely high impedances and reliable transmission is achieved by a pulse frequency modulation code in which the output of a neuron is transformed into an impulse frequency that is then integrated at the synapse. There are, however many examples such as in the retina where outputs from short axon neurons are continuous potential changes. Except in cases where very precise temporal relations must be preserved, as for example in the computation of acoustic delays, an individual impulse has little significance.

We have previously used pulsing neurons in earlier networks for acoustical pattern recognition [19] and have found no situation where they could not be replaced by neurons with continuous output. In fact when E_x is set high, the neurons described here will respond with phase locked pulses to sinusoidal inputs and can therefore also be used for acoustical delay computations. If in the future pulsing neurons appear essential we can easily insert modules containing such neurons into the machine.

The analog output multiplexer.

The multiplexer provides the host computer with time segments of the outputs from all neurons that are used for monitoring the network performance and as a basis for the implementation of learning algorithms and adjustment of synaptic weights and/ or connection architectures. It does not interfere with the actual operation of the net which is entirely in analog mode.

The multiplexer consists of 16 analog switches that connect the neuron outputs sequentially to a common output line, (see Fig. 3, 10 and 11). This output is buffered and provides a signal OM that is sent to an A/D converter over a common line. The output signals are stored in the memory of the host computer.

The switches of the multiplexer are addressed in series by complementary clock signals. The addressing circuit is a 16 bit shift register that shifts a "1" from the input to the output. The shift register is clocked by two phase clocks which are generated on chip from the master clock CK, coming from the host microprocessor. This clock runs at 2 MHz. After the last neuron has been read, the control circuit generates a pulse, ORO, that is sent to the ORI input of the next neuron chip. This module is now ready to send its 16 analog neuron outputs sequentially to the output line OM, after which it sends an ORO pulse to the next chip in the row. In this way all neurons are read out in sequence. When all chips have been read, the microprocessor sends an ORI pulse to the first neuron chip and the procedure starts over again. In order to synchronize the multiplexer with the microprocessor and the A/D converter, provisions are made for a control line phi2 between the neuron chip and the microprocessor.

By using one or more fast A/D boards (250KHz conversion frequency) the outputs of several hundred neurons can be read with msec time resolution which is fast enough to provide the host computer with time segments of the state of the network.

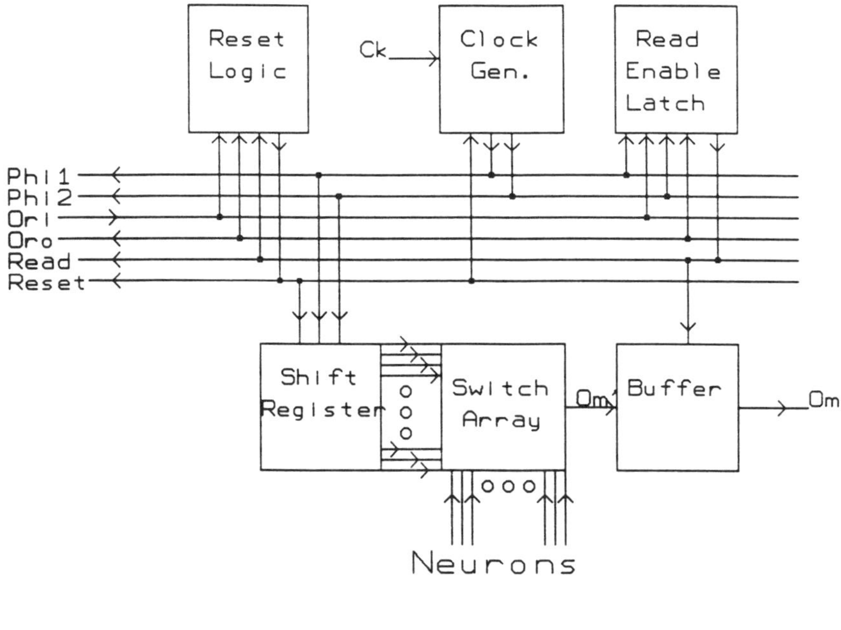

Fig. 10 Diagram of the analog multiplexer.

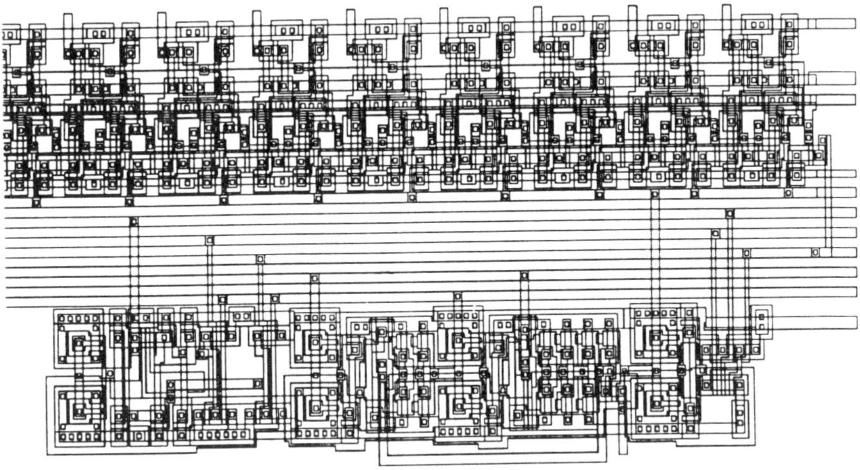

Fig. 11 Layout of the analog multiplexer. This design does not contain the buffer in Fig. 10. Only 8 of the 16 shift registers are shown. It has been fabricated but not yet tested.

THE SYNAPSE MODULE

Each synapse chip contains a 32 x 16 array of synapses. The synaptic weight of each synapse is set by serial input from the host computer and is stored at the synapse in a 6 bit local memory. The dynamic range of the synapse gains extends from 0 to 10 with 5 bit resolution, a sixth bit determines the sign. The gains are implemented by current mirrors that scale the neuron output after it has been converted from a voltage to a current.

The modifiable synapse designs reported in the literature use either analog or digital signals to set the gains.[2-5,9,25,26] For our initial design we chose the latter method because of its greater reproducibility and because direct analog setting of the gains from the neuron outputs would require a priori knowledge of and commitment to a particular learning algorithm. Nevertheless, since analog controlled synapses have the advantages of less area and continuous programmability compared to digital versions, we have fabricated but not yet tested an analog controlled synapse based on avalanche injection.

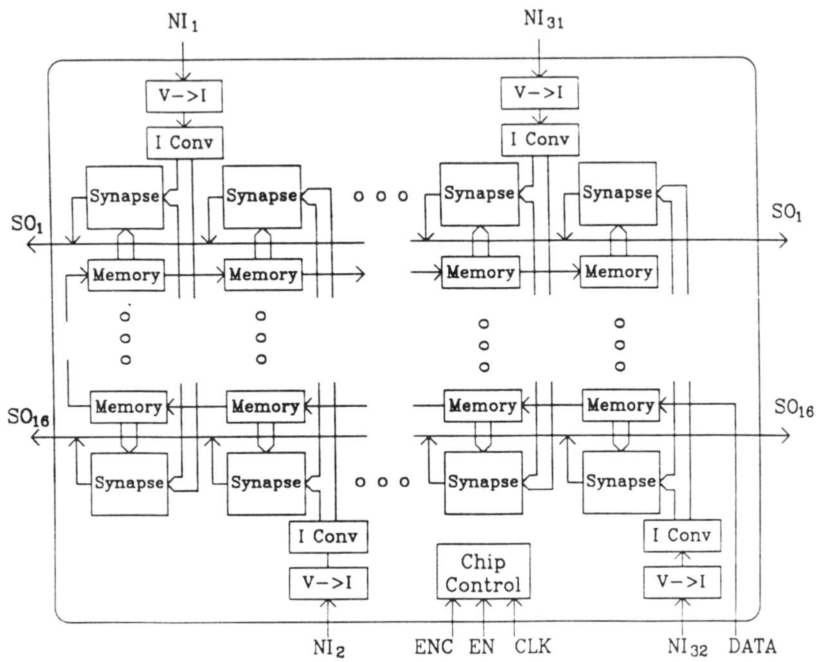

Fig. 12. Diagram of the synapse module. Each synapse gain is set by a 5 bit word stored in local memory. The memory is implemented as a quasi-dynamic shift register that reads the gain data during the programming phase. Voltage to current converters transform the neuron output (NI) into a current. I-Conv are current mirrors that scale the currents with 5 bit resolution. The weighted currents are summed on a common line to the neuron input (SO).

Layout and performance of the synapse module are shown in Figs. 12 - 16. The block diagram of the module is shown in Fig. 12. It consists of an array of 32 by 16 synapses and a similar array of 6 bit memory elements which is mapped onto the synapse matrix. The chip has 32 input lines (NIj) that are coming from neuron outputs, routed over switch modules. The inputs, which vary between 0 and 4 volts are transformed into a current by the V-I converter units shown at the top and bottom of the diagram. Associated with the converter is a current divider, which generates the required voltages to drive the synapse. Only one V-I converter and current divider is needed per column. There are 16 output lines, labeled SOj, which carry the sum of the current outputs of the 32 synapses on row i. These output lines are connected to the corresponding 16 inputs of the neighboring neurons (see overview Figure 1 and 2). Sixteen additional input lines (EIj) are provided in order to be able to increase the fan-in from 32 to 64 and up by placing one or more synaptic chips adjacent to each other and connecting the outputs (SO) of one chip to the extended inputs (EI) of the other.

The voltage to current converter takes the neuron output voltage and generates a current proportional to the voltage. The circuit, derived from a design in reference 21, is shown in Fig. 12. Associated with the converter is a current divider based on a series of current mirrors. This circuit generates currents which decrease in a logarithmic fashion allowing the selection of synaptic gains (weights) over a range of 0 to 10 with a 5 bit resolution.

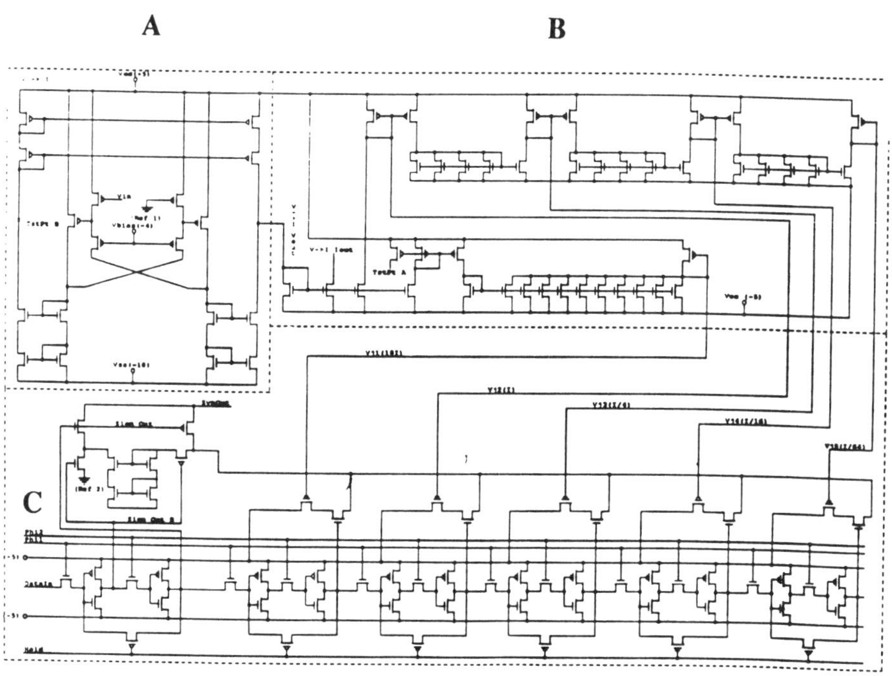

Fig. 13. The voltage to current converter and the synapse. The V to I converter is shown in **A**. The current divider (**B**) consists of current mirrors in series. The synapse (**C**) is a steered current circuit. The currents in the PMOS transistors are derived from the current divider circuit. A current switch steers the current to the neuron input line or to ground. The switch is controlled by the memory (**C**). A current mirror inverts the current to implement excitatory or inhibitory connections.

The synapse itself consists of a series of current switches in series with a current source, schematically shown in Fig. 13. The current source is realized by a transistor whose gate and source terminals are connected to the current divider circuit, in a current mirror configuration. Separating the current divider circuits from the synapse allows sharing the divider circuit among all the synapses on the same column. A current inverter controlled by the sign bit allows implementation of excitatory and inhibitory connections without doubling the number of inputs. The switches that select the current levels and thus determine the synaptic weights are driven by the outputs of the memory elements, as shown in Fig. 13.

The memory elements consist of cross coupled inverters which are connected in series to form one large shift register. This allows the use of the same inverters to read the data serially in all memory elements by using a two phase clocking scheme. The data is provided by the digital host computer over one line, labeled DATA in Fig. 12. The layout of the synapse components and a photograph of a prototype chip are shown in figs.14 and 15.

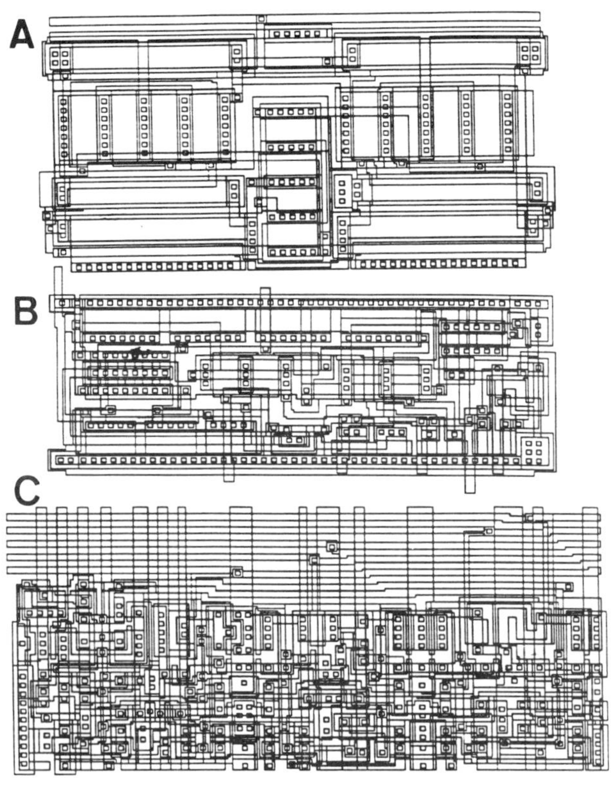

Fig.14. Layouts of the synapse components showing in **A** the V-I converter, in **B** the current divider and in **C** the current selector and memory.

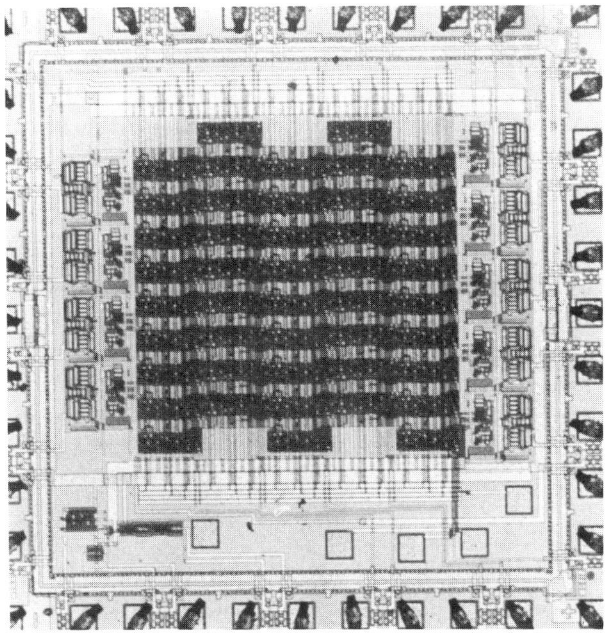

Fig. 15. Photograph of a prototype Synapse chip containing an 8x5 array of synapses. The chip was fabricated in 2u CMOS.

As seen in Fig. 16A, the synaptic transfer function is linear from 0 to 4 V. Fig. 16B shows that the synapse exhibits less than 10% variation from chip to chip.

The use of current mirrors permits arbitrary scaling of the synaptic gains (weights) with trade off between range and resolution limited to 5 bits. Our current design calls for a minimum gain of 1/64 and a maximum of 10. The lower end of the dynamic range is determined by the number of possible inputs per neuron which when active should not drive the neuron output to its limit, whereas the high gain values are needed in situations where a single or only a few synapses must be effective such as in the copying of activity from one neuron to another or for veto inhibition. The digital nature of the synaptic gain control does not allow straightforward implementation of a logarithmic gain scale. Fig. 17 shows two possible relations between digital code and synaptic gain. In the first case, implemented in our current design, the total gain is the sum of 5 individual gains each controlled by one bit. This leads inevitably to jumps in the gain curve. In a second case which we are currently investigating, a linear 3 bit gain is multiplied by four different constants controlled by the 4th and 5th bit. This "floating point" scheme provides better approximation to a logarithmic scale. The synapse specifications are summarized in Table 2.

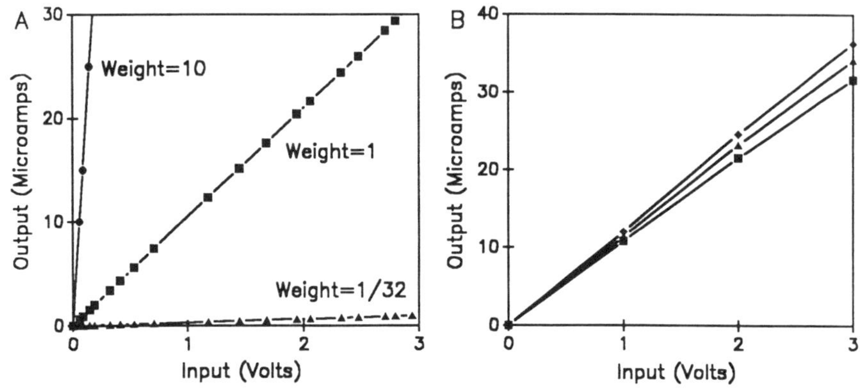

Fig. 16. **A.** synapse transfer characteristics for three different settings. The data were obtained from the chip shown in Fig. 13. **B.** transfer characteristics obtained from four different test chips.

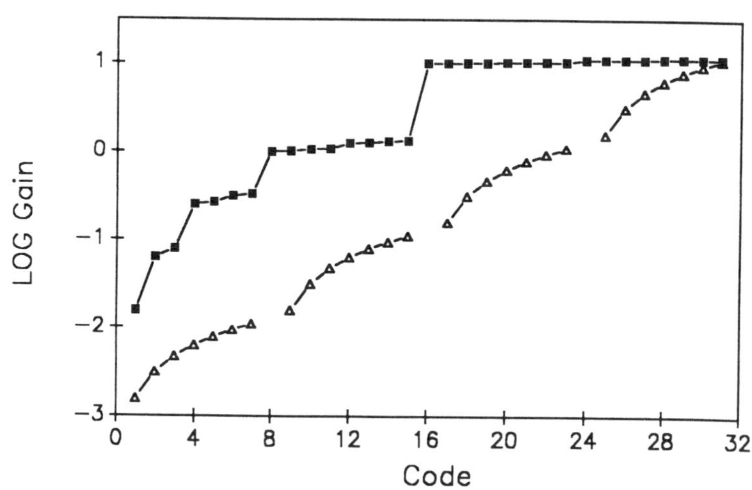

Fig 17. Digital Code vs. synaptic gain, squares are current design, triangles represent a five bit floating point format.

Although the resolution of an individual synapse is limited to 5 bits, several synapses driven by one neuron can be combined through switching, permitting greater resolution and dynamic range. Furthermore, mismatching of synaptic currents due to transistor differences can be compensated by this method.

It might seem that the limited number of inputs per neuron restricts the computations that can be performed by any one neuron. However the results obtained by one neuron can be copied through a unity gain synapse to another neuron which receives the appropriate additional inputs.

TABLE 2

SYNAPSE SPECIFICATIONS

Process	2u CMOS
Operation	Current Scaling
Weight control	Digital, 6 bits
Dynamic range of weights (gain)	0 to + - 10
Resolution	integer 5 bits + 1 sign bit[1] floating point 3 + 2 bits[2]
Output Current Range	0 to 400 uA
Transfer Characteristic	Linear from 0 to 4 V
Download time	3us /synapse
Input Impedance	> 10^{12} Ohm
Number of Synapses per Chip	514
Number of Input Lines	32

1) Currently implemented, 2) Under development.

THE SWITCH MODULE

The switch modules serve to route the signals between neuron modules and the synapse modules thereby changing the connection architecture. A similar routing scheme has been employed to make programmable interconnections between subcircuits on a VLSI chip.[30] Each module contains a 32 x 32 cross point array of analog switches which are set by serial digital input. Figs. 18 and 19 show a block diagram of the chip's major subsections. They consist of switching fabric, control logic, serial-in parallel-out shift register (SIPO) and control, write strobe generator (XGen) and gated two-phase clock generator (2PG).

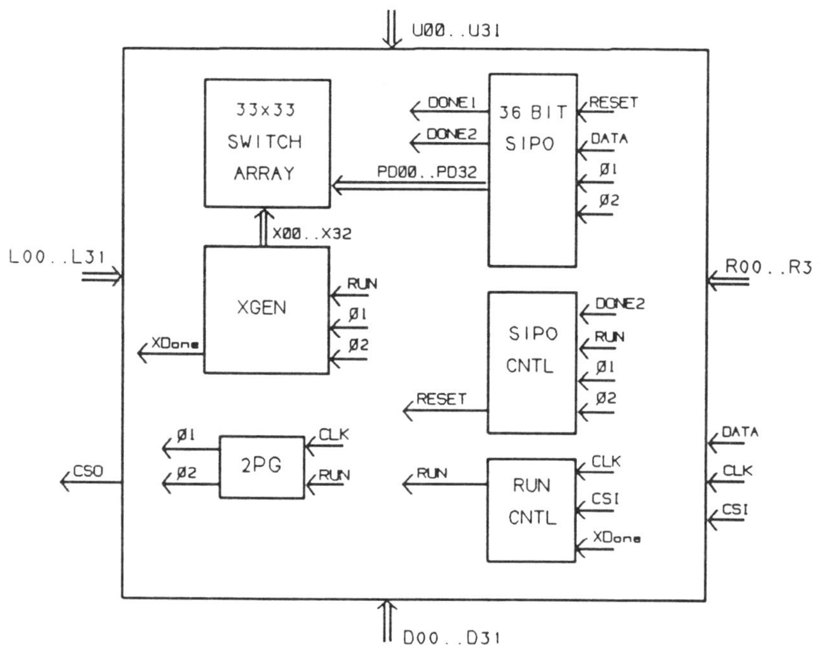

Fig. 18. Block diagram of the switch chip. Data is downloaded serially from the host computer, parallelized by the SIPO logic, and transferred into the switch array by the XGEN circuit.

Fig. 19 shows a block diagram of the switching fabric. The signals U00..U31 pass vertically through the chip to D00..D31; similarly the signals L00..L31 pass horizontally through to R00..R31. Each square represents a one-bit switch control memory and analog switch. The control data can connect an arbitrary horizontal line to a vertical line by writing a '1' into the appropriate memory cell.

The circles along the right and bottom edges also represent switches and memory. The switches that are in series with the horizontal or vertical signals allow the microprocessor to "cut" a horizontal or vertical trace in the switch chip. This allows the interconnection buses to be partitioned into several segments which increases the maximum number of obtainable connection. In addition to switches the modules contain circuits which control the time constants of the synapse transfer function (see figs. 19, 22 and 23).

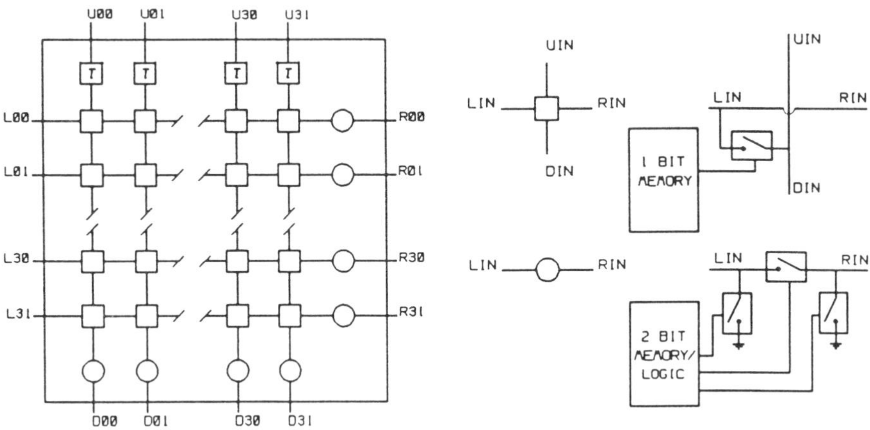

Fig. 19. Diagram of switching fabric. Squares and circles represent switch cells which connect the horizontal and vertical connectors or cut the conductors. The units labeled T represent programmable time constants.

Each switch cell is implemented as a CMOS transmission gate connected to one bit of memory. The control logic subsystem enables chip load circuits when CSI input is active, disables chip load circuits when loading is done and propagates CSO to the next switch chip. The global control signal RUN is asserted while the chip is loading data. The switch performance is summarized in Table 3.

The SIPO parallelizes serial data received from the microprocessor into 33 bit words. Each SIPO bit (PD00..PD32) drives one row of switch memories. Additionally, several SIPO taps are used to generate the control signals SIPODone1 and SIPODone2. These signals are used to detect when the first 33 bits of a cycle have been received from the microprocessor and to count the 36 clocks that comprise one cycle.

The XGen circuit (Fig. 18) maintains the address of the next column to receive SIPO data. After a 33 bit word has been assembled by the SIPO, XGen writes it to the proper column of switch memory by asserting one of the X00..X32 control lines. This function is implemented by shifting a "1" through a 33 bit shift register. The XDone output is generated after all 33 columns have been loaded; this is used by the Control logic to generate the CSO and to shutdown the chip.

The gated two phase clock generator (2PG) produces the non-overlapping two phase clock. The 2PG block includes logic which enables the clock only while the microprocessor is loading the switch memory. This reduces the chip's power consumption.

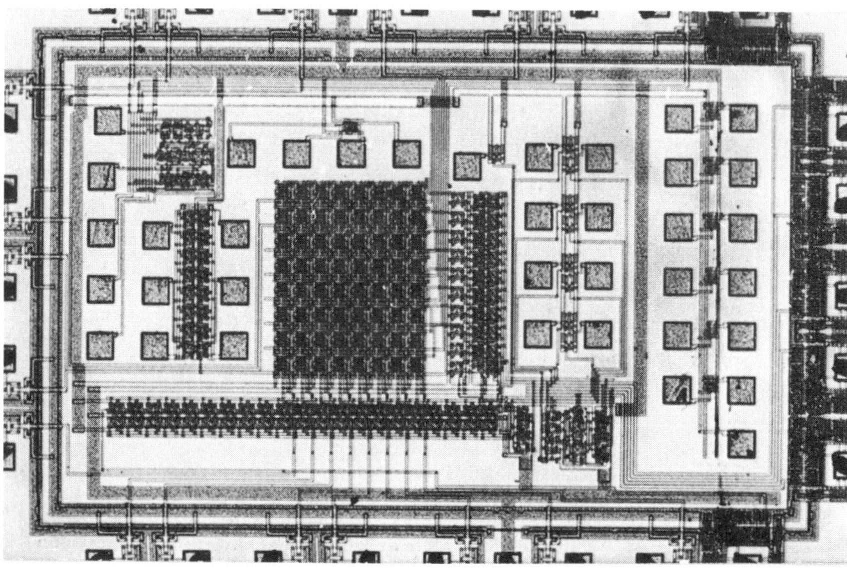

Fig. 20. Photograph of a switch module test chip.

Figure 21 illustrates the microprocessor interface timing diagram. CLK is a system-wide 2 MHz, 50% duty cycle clock. CSI and CSO are daisy chained signals used to select a single Switch Chip for loading. DIN is a bused signal used to serially transfer data from the CPU to the Switch Chips.

The load operation is initiated when the Switch Chip detects CSI active during the falling edge of CLK. CSI is driven by the microprocessor or by the CSO output of the previous switch chip. Thirty-three cycles (with 36 CLKs/cycle) are used to serially transfer data into the switch memory. On the 36th CLK of the 33rd cycle, the Switch Chip generates the CSO signal. This informs the next Switch Chip to commence loading. The CSO output of the last cascaded Switch Chip is looped back to the processor allowing the CPU to determine the number of Switch Chips in the system at run-time.

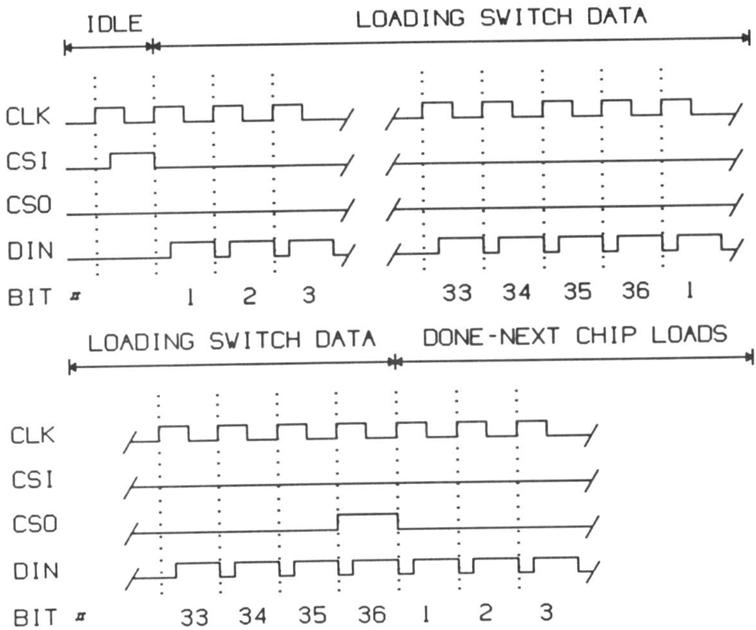

Fig. 21. Timing Diagram for loading the switch memory.

TABLE 3

SWITCH CHIP PERFORMANCE

Process	3u CMOS
Input capacitance	1pF
On resistance	< 3 KOhm
Download time	0.5 us/switch
Off resistance	> 1 TOhm
Memory/switch size	75u x 90u

During each 36 CLK cycle, 33 bits of switch control memory are loaded. The data provided by the processor during the last three CLK's of a cycle are not used; this allow the CPU to nybble-align the switch data in its local memory. The amount of memory used by the CPU to maintain the image of the switch memory is 33 x 36 = 1188 bits. This means that approximately 35 KBytes are necessary in a system consisting of 12 Switch Chips/row x 20 Rows/Card = 240 Switch Chips.

We chose a control scheme for the switch chip that differs from that of the synapse chip in order to evaluate their relative merits. The control logic used for the synapse chip has certain advantages and will be incorporated into the next version of the switch chip.

ADJUSTMENT OF SYNAPTIC TIME CONSTANTS

For the analysis or generation of temporal patterns as they occur in motion or speech, adjustable time constants of synaptic transfer must be available. This is a very important aspect of neural computation and is only beginning to be recognized. For a discussion of this topic see refs. 18, 19, 22, 23, 24. From our experience, low pass filtering of the input signal to the synapse with 4 bit control of the time constant over a range of 50 us to 500 ms seems sufficient to deal with real world data. By combining the low passed input with a direct input of opposite sign, both originating from the same neuron, the typical "ON" and "OFF" responses which serve as measures of time after beginning and end of events and are common in biological systems can be obtained (see fig. 22).

Several methods to implement large and adjustable time constants are being investigated. One way is to charge or discharge a capacitor (of a few pF) through a transconductance amplifier, connected as a unity gain buffer.[1] If one biases the amplifier with currents in the subnanoampere range one can obtain a very low gain-bandwidth (GBW) and slew rate. When an input step is applied to the amplifier the output will first slew linearly and then evolve exponentially towards its final value with a rate proportional to the GBW. By adjusting the bias current of the amplifier one can in effect change the time constant of the circuit. This circuit has been simulated and found to perform creditably (see fig. 23).

Another scheme is based on translinear circuits. By biasing MOS transistors into weak inversion, the voltage-current characteristics is exponential similar as in a bipolar transistor. This permits the use of the flexibility of bipolar transistors to implement different functions. One possibility is to generate very small currents by using a (variable) resistor in series with a transistor in one branch of a current mirror. This current can now be used to charge or discharge a capacitor over tens and hundreds of milliseconds. This circuit can also be used in combination with a source follower to obtain very small transconductances and hence large time constants. The choice among these and other schemes will be based on area, power consumption, ease of implementation and uniformity of the time constants over a chip and from chip to chip.

The low pass circuits will be placed on the switch chips at the points shown in Fig 19. Since not all synapse inputs need to have this feature, the circuit will be placed on only a limited number of lines on the switch chip.

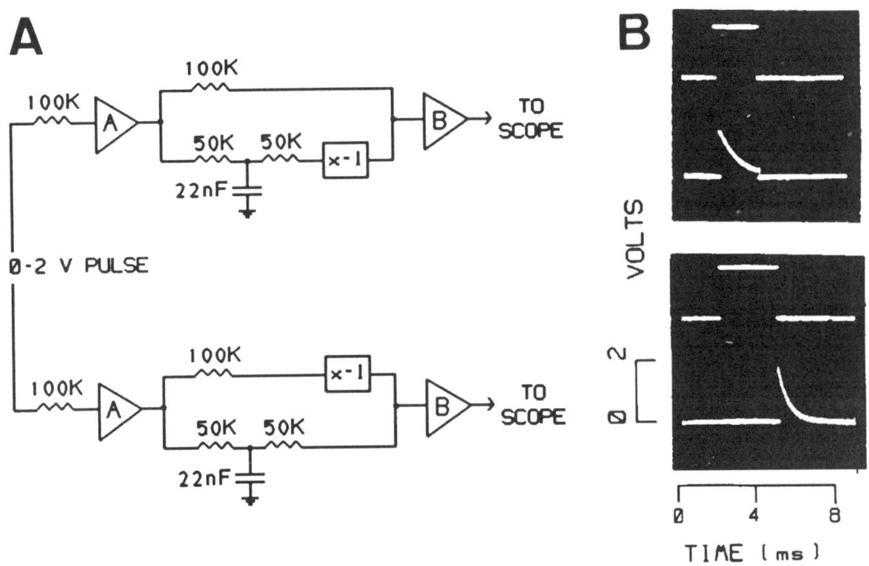

Fig. 22 **A.** Circuit consisting of neurons, synapses, and and low pass synaptic tranfer functions to obtain "ON" and "OFF" responses. **B.** Waveforms recorded from the output neurons (bottom trace) in response to a step input (top trace).

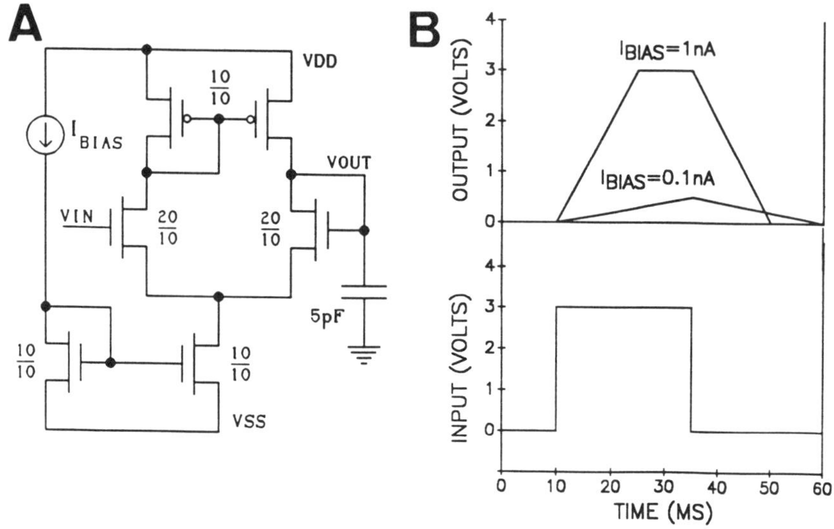

Fig. 23 A. Circuit for obtaining variable synaptic time constants. By using very low bias currents (I_{Bias}) in a transconductance amplifier very small slew rates can be obtained. The simulated response of this circuit to a square wave is shown in **B.** Although these responses are not exponential, the frequency response of the circuit approximates that of a low pass filter.

INTEGRATION OF THE COMPONENTS

Figs. 24 A to D demonstrates that the synapses, switches and neurons operate together as specified. A prototype switch chip, synapse chip and neuron chip were interconnected as shown in fig. 24A. Switch settings and synapse weights were controlled by software from a digital host.

In fig. 24B the weight (gain) of a single synapse was set in steps from 1/32 to 1.8. In fig. 24C two different inputs were selected on the switch chip and summed at the neuron input. Fig. 24D demonstrates the effect of increasing the gain of an inhibitory synapse on neuron output.

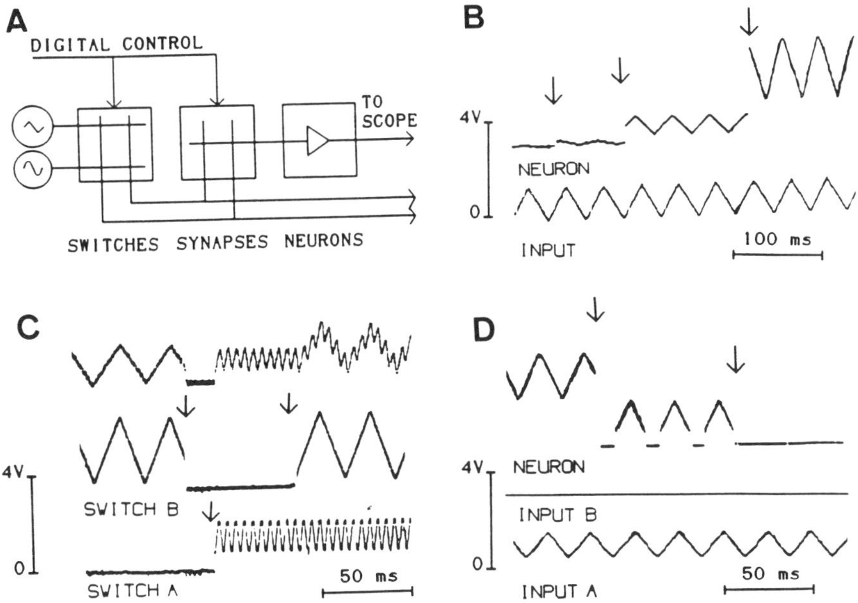

Fig. 24 **A**. Schematic of the interconnections between three different module chips. Input signals were from function generators, outputs were recorded either from the neuron output or at the switch outputs. The switch and synapse settings were controlled from a digital host computer. The chips were prototypes as shown in figs. 8, 15, 20.

B. Control of neuron output through adjustment of synaptic gain. At the times indicated by the arrows the gain was switched in sequence from 1/32 to 1/8, 1/2 and 1.8.

C. Two different inputs are selected through the switch chip either alone or combined. The inputs to the two synapses are recorded at the output from the switch chip (bottom records). The neuron output is the top trace.

D. This figure shows the suppression of neuron output by inhibition with a DC input through a second synapse. Input A to the excitatory synapse was a triangular wave plus a constant voltage. Input B to the inhibitory synapse was a constant voltage. Increasing the gain of the inhibitory synapse increases the inhibition. The mode (excitation or inhibition) was selected by switching the sign bit of the synapse. Notice the step at threshold which results from setting the extra output at threshold (Ex) to 0.6 V.

LOGIC CONTROL

The switches, synaptic weights and neuron parameters are set by serial input from the host computer.

In our current design groups of the different chips are connected and loaded as a daisy chain where the chips are sequentially enabled and the uploading continues until all chips are loaded. This method, discussed above has the disadvantage that all chips in a group must be reloaded even when changing only a single parameter on one chip.

We are considering a simpler method, where each chip could be addressed and uploaded separately. This is be achieved by adding a single memory unit to each chip. These memory units are connected from chip to chip and form an elongated shift register similar in design to the memory of the switches and synapses. The contents of this memory enable the chip to upload data from a separate bus. The host computer would shift a single bit into the enabling shift register and then load the data into the enabled chip. This method has the added advantage that identical data can be loaded into different chips in parallel.

With a clock rate of 2 MHz data upload times are currently 1.2 ms per switch chip and 3.3 ms per synapse chip. Higher clock rates are feasible and would reduce upload times accordingly.

PACKAGING

The smaller chips containing test structures and individual components are packaged by MOSIS in standard 40 pin dips. The complete modules will be packaged in 160 lead surface mount quad packs (Fine Pitch EIAJ Standard). Input and output lines are arranged at right angles with identical leads on opposite sides, as shown in fig.25. The packages are soldered by tape directly on circuit boards that provide the interconnections between modules.

In an alternative method providing higher pin out, the chips are mounted on planar chip carriers that are connected by elastomeric connectors. This design allows easy replacement of individual modules but the connections may be less reliable than with the surface mount packages. However, elastomeric connectors are used routinely in watches where they provide reliable service.

More sophisticated connection techniques such as flip chip mounting or deposition of metal lines between closely opposed chips are also possible.

The chips were designed in a 3 and 2 u CMOS technology. At 0.9 u the number of neurons or synapses per chip could be doubled or neurons and synapses could be integrated on one chip thereby reducing the need for off chip connections.

We plan to assemble a prototype machine containing 64 neuron modules for a total of 1024 neurons together with appropriate numbers of switch and synapse modules on one board. Nets larger than this would require connections between boards that could be made by plastic tape with deposited metal lines. The latter method shall be used also to connect the digital control circuits to the host computer.

Fig. 25. Schematic drawing of a neuron chip mounted on a chip carrier. The carrier is 2 cm wide and has 40 contacts on each side. The horizontal continuous lines carry control signals.

SOFTWARE CONTROL AND OPERATION

Connections, synaptic gains, neuron parameters and time constants are set from the host computer either manually or through implementation of learning algorithms that derive these parameters on the basis of the neuron outputs. The connections would be routed under graphic control or through routing routines as they are used in circuit board design. Eventually we envision developing a macro language that would generate and store libraries of computational architectures which could be linked into larger systems for specific tasks. Examples of such subsystems are feature specific receptor fields,[27-29] temporal pattern analyzers, or circuits for motion control.

The primary areas of application include real-world, real-time or compressed-time pattern analysis and recognition, robotics, the design of dedicated neural circuits and the exploration of different learning algorithms. Input to the machine can come from sensory transducer arrays such as an electronic retina, cochlea [8,15,16] or tactile sensors. For other computational tasks, input is provided by the host computer through activation of selected neuron populations via threshold control.

The learning speed depends very much on the algorithm. In dynamic cases, where execution mode becomes rate limiting, the gradient descent methods such as backpropagation would be executed faster than on digital machines. Feed-forward algorithms such as sequential convolutions can be executed in a few cycles and would be very efficient.

The full potential of the machine is realized in execution mode, especially in situations involving neural computation of dynamic systems i.e. situations in which time is a variable. Acoustical pattern recognition and the computation of moving images are prominent examples. But even for rapid recognition of stationary patterns the machine would prove superior to sequential machines. High speed character recognition, sorting of particle tracks in high energy physics or target acquisition from fast moving vehicles fall into this category. In general, systems that require solutions of many simultaneous equations would benefit. A large scale winner-take-all computation where the units are mutually inhibiting is extremely time consuming for digital methods but can be solved in real time by analog computation.

In these applications the machine could exceed by orders of magnitude the computational speed of any currently available digital computer. An estimate of attainable speed can be made as follows:

Consider the network shown in fig. 26 with N neurons each receiving M inputs. Each input has an RC stage generating time constants in the range of 1ms to 1 sec and a gain stage (G_{ij}) which establishes the input's weight. The weighted sum of a neuron's inputs is fed into an amplifier which has a sigmoidal transfer function.

The network is described by N*M differential equations which yield the voltages on the capacitors as a function of time and the neuron outputs.

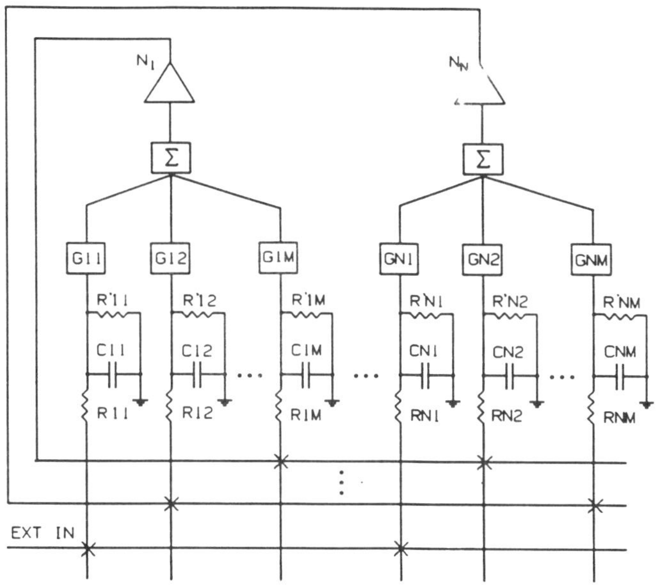

Fig. 26. Electrical model of the simulated network.

In the above figure the voltage V_{ij} (the voltage on the jth capacitor of the ith neuron) is given by:

$$dV_{ij}/dt \quad = \quad -V_{ij}/C_{ij} * (1/R_{ij} + 1/R'_{ij}) + N_x/C_{ij}*R_{ij} \qquad (1)$$

where N_x is the voltage from the neuron driving this input

The output voltage of the jth neuron is given by:

$$N_j \quad = \quad S\left(\sum_{i=1}^{M} V_{ij}*G_{ij}\right) \qquad (2)$$

where $S(x) = 1 / (1 + \exp(-x))$

These equations were solved with a fourth order Runge-Kutta algorithm using an adaptive step size [31]. The gains G_{ij} were uniformly distributed in [-1..1] while the interconnections were randomly selected. The external input was a square wave whose period was 5ms and duty cycle was 50%. The equations were integrated for 5ms of simulated time on a Sun 4/110 workstation with a floating point unit.

Table 4 presents the CPU time used to solve various sized networks with randomly selected time constants in the range of 1ms to 1 sec. Each entry represents the average of at least 20 random networks and time constants.

TABLE 4

SIMULATION RESULTS

Neurons	Inputs/Neuron	CPU time for 5 ms simulation (seconds)
16	8	8.9
16	12	18.9
16	16	25.9
32	8	26.7
32	12	55.6
32	16	104.6

The data in table 4 indicate that the CPU time scales greater than O(N*M). However, even if we make we the conservative assumption of linear scaling, the results demonstrate that a Sun 4/110 rated at 5 MFLOPS requires 20-40 seconds of CPU time per connection to integrate one second of network time. The neural analog machine configured for 1000 neurons each with 100 inputs could therefore run at speeds equivalent to more than 10^{11} FLOPS. Larger networks would scale accordingly.

The analog hardware can easily support 1 usec time constants which would increase the speed advantage by another factor of 10^3. Finally, the neuron transfer function assumed for the simulations was a sigmoid which is differentiable everywhere. The VLSI hardware also supports a thresholding transfer function (see fig. 9) which has a finite number of discontinuities in its derivative. Simulation of this transfer function requires extremely small steps, increasing simulation times 100 fold.

Acknowledgement

The research described in this paper was supported by grants from the Office Of Naval Research (N00014-89-J-1249) and the National Science Foundation (EET87-166685)

REFERENCES

[1] Mead, C.A. Analog VLSI and Neural Systems. Addison-Wesley, 1989

[2] Raffel, J.I., Mann, J.R., Berger, R., Soares, A.M., Gilbert, S. A Generic Architecture for Wafer-Scale Neuromorphic Systems. **IEEE First International Conference on Neural Networks**, III-501, San Diego, CA 1987.

[3] Alspector, J., Allen, R.B. A Neuromorphic VLSI Learning System. Advanced Research in VLSI. **Proceedings of the 1987 Stanford Conference**, 1987.

[4] Tsividis, Y., Satyanarayana, S. Analogue Circuits For Variable-Synapse Electronic Neural Networks. **Electronics Letters** 23:1313-1314, 1987

[5] Schwartz, D., Howard, R., Hubbard, W. A Programmable Analog Neural Network Chip, **IEEE J. of Solid State Circuits**, (to be published).

[6] Graf, H.P. et al. VLSI Implementation of a Neural Network Memory with Several Hundreds of Neurons, **Neural Networks for Computing, AIP Conference Proceedings,** 151:182-187, 1986.

[7] Mueller, P., Van der Spiegel, J., Blackman,D., Chiu, T., Clare,T., Dao,J., Donham,,C., Hsieh,T. Loinaz,M. Programmable Analog Neural Computer and Simulator,In **Advances in Neural Information Processing Systems**,Proceedings of the IEEE Conference Denver 1989, D. Touretzky Ed., Morgan Kaufmann , San Mateo, CA.

[8] Chiang,A., CCD Retina and Neural Network Processor, **Hardware Implementation of Neural Nets and Synapses**, Report on a Workshop. P. Mueller ed. San Diego 1988.

[9] Shoemaker,P.A., Shimabukuro,R. A Modifiable Weight Circuit for Use in Adaptive Neuromorphic Networks. **Neural Networks**, 1,Suppl.I, 409, 1988.

[10] Sage, J.P., Thompson,K.,Withers, R.S., An artificial Neural Network Integrated Circuit Based on MNOS/ CCD Principles, **Proc. Conf. Neural Networks for Computing, AIP** 151, 381 1986.

[11] Rudnick,M., Hammerstrom,D., An Interconnect Structure for Waferscale Neurocomputers, **Neural Networks**, 1,Suppl.I, 404, 1988.

[12] White, J.,Furman, B., Abidi, A.A., Baker, R.L., Mathur, B., Wang, H.T., Parallel Architecture for 2D Gaussian Convolution of Images, **Neural Networks**, 1, Suppl. I, 415, 1988

[13] Furman, B.,White, J.,Abidi, A.A., CMOS Analog IC Implementing the Back Propagation Algorithm, **Neural Networks**, 1,Suppl.I, 381, 1988.

[14] Akers, L.A., Walker, M.R., Training a Limited Interconnect Feedforward Neural Array, **Neural Networks, 1, suppl.I 381,1988.**

[15] Mead, C. and Mahowald, M.A. A Silicon Model of Early Visual Processing. **Neural Networks**, 1:91-97, 1988.

[16] Lyon, R.F. and Mead, C.A. An Analog Electronic Cochlea. **IEEE Trans. Acoust., Speech, Signal Processing**, 36:1119-1134, 1988.

[17] Lazarro, J.P., Ryckebusch, S., Mahowald, M.A., and Mead,C.A., Winner Take All Networks of O(n) Complexity, **Proc.IEEE Conference on Neural Information Processing Systems**, Denver, CO, 1988.

[18] Mueller, P., Lazzaro, J. A Machine for Neural Computation of Acoustical Patterns.**Neural Networks for Computing, AIP Conference Proceedings**, 151:321-326, 1986.

[19] Mueller, P., Martin, T., Putzrath, F., General Principles of Operation in Neuron Nets with Application to Acoustical Pattern Recognition. **Biological Prototypes and Synthetic Systems**, E.E. Bernard and M.R. Kare Edt., Plenum Press New York. 1962.

[20] Rumelhart, D.E., Hinton, G.E., Williams, R.J., Learning Internal Representations by Error Propagation, **Parallel Distributed Processing V. 1,** 319 **MIT Press** 1986.

[21] Bult, K. and Wallinga, H. A Class of Analog CMOS Circuits Based on the Square Law Characteristics of an MOS Transistor in Saturation. **IEEE J. Solid-State Circuits**, SC-22:3657, 1987.

[22] Mueller P., Principles of Temporal Pattern Recognition in Artificial Neuron Nets, **Artificial Intelligence**, S 142 , The Institute of Electrical and Electronic Engineers, Inc., New York 1963.

[23] Mueller, P., Computation of Temporal Pattern Primitives in a Neural Net for Speech Recognition, In. Report on a Workshop **"Hardware Implementation of Neuron Nets and Synapses"**, P. Mueller Ed., San Diego CA., 1988..

[24] Mueller, P. Computation of Temporal Pattern Primitives in a Neural Net for Speech Recognition. **Abstracts of the First Annual INNS Meeting**, Boston, MA, 1(1):308, 1988.

[25] Hu,V., Kramer, A., Ko,P.K., EEPROMS as Analog Storage Devices for Neural Nets, **Neural Networks,1 Suppl. I** 385 , 1988.

[26] Thakoor, A., Modifiable Synapses,In Report on a workshop **"Hardware Implementation of Neuron Nets and Synapses"**, P. Mueller Ed., San Diego CA., 1988..

[27] Mueller, P., Blackman,D., Spiro,P., Furman,R.E.,Neural computation of visual images, **IEEE First Annual Conference on Neural Networks**, 4 , 75, San Diego CA, 1987.

[28] Mueller,P.,Blackman, D.,Furman,R.E., Neural computation of Visual Images,In : **An Introduction to Neural and Electronic Networks**, S.F. Zornetzer, ed., Academic Press, 1989.

[29] Mueller,P., Neural computation of Pattern Primitives, **AIP Conference on Neuron Networks** Snowbird Utah 1988.

[30] Sivilotti, M., A Dynamically Configurable Architecture for Prototyping Analog Circuits, **Advanced Research in VLSI, Proc. of the Fifth MIT Conference.** Allen and Leighton eds., MIT Press, Cambridge MA.1988.

[31] Press, W.H., Flannery, B.P., Teukolsky, S.A., Vetterling,W.T., **Numerical Recipes,** Chapter 15,Cambridge University Press, 1986.

7

A CHIP THAT FOCUSES AN IMAGE ON ITSELF

T. Delbrück
California Institute of Technology
Pasadena, California, 91125
e-mail: tobi@hobiecat.caltech.edu

In the modeling of neural systems, time is often treated as a *sequencer*, rather than as an *expresser* of information. We believe that this point of view is restricted, and that in biological neural systems, time is used throughout as one of the fundamental representational dimensions. We have developed this conviction partially because we model neural circuitry in analog VLSI, where time is a natural dimension to work with, and we believe there are deep similarities between the technology we use and the one nature has chosen for us.

Neurobiologists are beginning to explore neural control systems that self-generate sensory input. The focus chip we report here models the focusing system of our eye. The human focusing mechanism is a one-dimensional control system in which experimenters have access to both visual input and motor output signals. For our model, the primary hypothesis about this system is that control signals are *generated actively*, by the motor system in the course of control. We have built and partially characterized a model system, using analog VLSI circuit primitives already developed for other purposes, that incorporates this hypothesis. This chip focuses an image on itself, using time domain information about the quality of the optical image and the motion of the lens.

THE HUMAN ACCOMMODATION SYSTEM

The process by which the eye focuses an image onto the retina is called *accommodation*. The eye accommodates by distorting the curvature of the lens. When muscle fibers running radially outward from the lens contract, the increased tension on the lens flattens it, focusing farther away. When muscle fibers running circumferentially around the lens contract, the decreased tension on the lens allows it to bulge, focusing closer (Weale, 1960). In our model system the focus is changed by changing the distance between a rigid lens and the chip, much like focusing a camera.

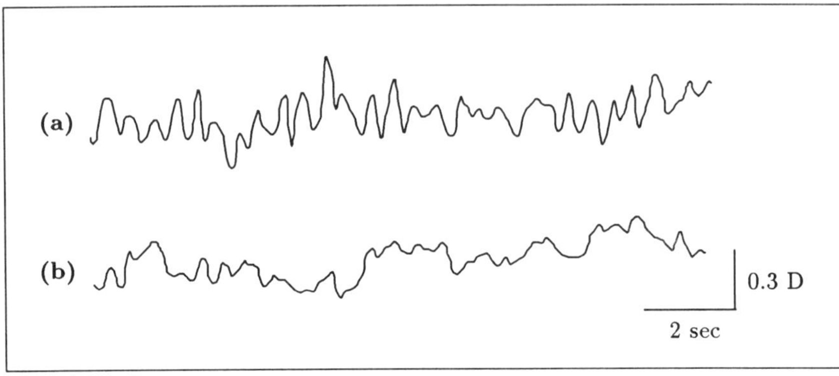

Figure 1. Recordings of fluctuations in human accommodation. These records were obtained using an infrared split-beam optometer. The optical distance to the target was 1 D. **(a)** Pupillary aperture was 7 mm. **(b)** Pupillary aperture was 1 mm. The retinal illumination was kept the same for each trial. (Adapted from Campbell *et al.* (1959) with permission.)

The stimulus for accommodation is not known, although there are results that positively indicate certain possibilities. The obvious possibility is the blur of the retinal image. The problem with imagining a control system for accommodation that uses retinal blur is that blur is an even-error signal: static blur does not say which way to change the accommodation to sharpen the retinal image. Other possibilities for the error signal that are odd-error have been proposed. In an elegant set of experiments, Campbell and Westheimer (1959) showed that chromatic and spherical aberration were sufficient odd-error signals in subjects with paralyzed accommodation. They did not show that these were necessary cues in subjects with normal accommodation reflexes. Given the possibility that image blur is one of the primary cues to accommodation, we might ask, exactly what functional of the image is used as the primary cue? The precise answer is unknown, though there are clues. For example, Fujii *et al.* (1970) showed that intensity gradient was more important than total contrast modulation, as a stimulus to accommodation.

Accommodation fluctuates even under steady-state conditions. The existence of these fluctuations has been known, or at least postulated, for a long time (see, for example, Helmholtz, 1924). Campbell *et al.* (1959) were among the first researchers to obtain recordings of these fluctuations. Figure 1 shows examples of these fluctuations; Figure 2 shows power spectra for the records in Figure 1. We can see that the amplitude of the fluctuations decreases with a smaller pupillary aperture, and their character changes. The fluctuations have a characteristic frequency that shows up as a pronounced bump in the power

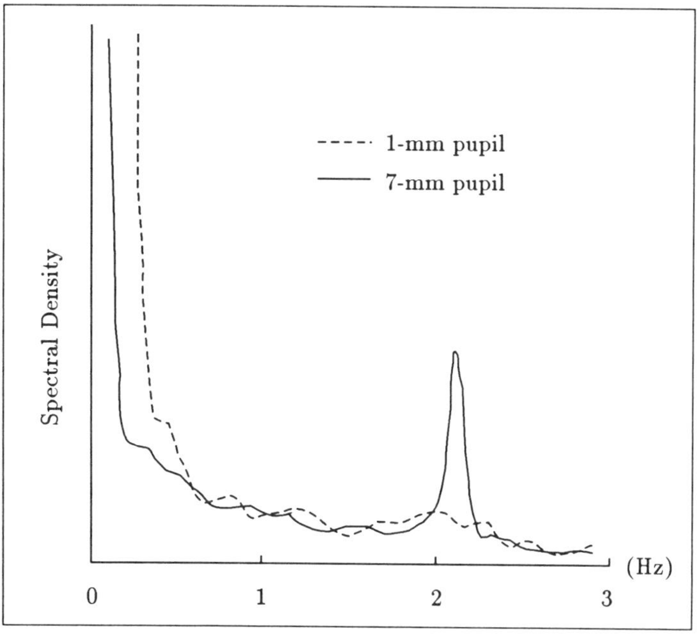

Figure 2. Power spectra for the records from Figure 1. Note that the peak in the power spectrum for the signal in Figure 1(a) disappears in the spectrum for Figure 1(b). (Adapted from Campbell *et al.* (1959) with permission.)

spectrum, usually near 2 Hz. The amplitude of the fluctuations increases as the optical distance to the target gets smaller (Denieul, 1982). For optical distances of 4 D (=25 cm viewing distance), the size of the fluctuations can grow larger than 0.1 D ($\approx \pm$ 0.5 cm fluctuation in focal distance). These oscillations are below perceptual thresholds under ordinary conditions, yet stimuli oscillating at below the perceptual threshold can drive the accommodation system (Kotulak and Schor, 1986b).

THE MODEL OF ACCOMMODATION USED BY THE CHIP

In our model, the measure of image quality is denoted by the term *sharpness*, or by the symbol s. The state of accommodation is denoted by the symbol l. In the current physical realization of our system, the lens is moved, rather than distorted, so this accommodative state is equivalent to the lens position, relative to the point at which the lens settles in the absence of any stimulus. The idea for our circuit came from a paper by Kotulak and Schor (1986). The essence of the idea is simply stated: The sign of $\dot{s}$ † indicates whether accommodation is changing in the correct direction, and the sign of $\dot{l}$ indicates in which direction the accommodation is currently changing. The sign of the product $\dot{s}\dot{l}$ gives the *sign* of the error signal for $\dot{l}$.

Still open is the question of what to use as the error *magnitude*. Kotulak and Schor (1986) suggested that $|\dot{s}/\dot{l}|$ would be a reasonable choice; for the model reported here, $\tanh(\dot{s}\dot{l})$ is used as the error signal for $\dot{l}$. We integrate this error signal with respect to time, using the mass of the lens. The discussion section of this report notes some other possible uses of the error signal, besides integrating it with a mass.

We can now write a dynamical equation for the motion of the lens:

$$M\frac{d}{dt}(\dot{l}) = S\tanh\left(\dot{s}\dot{l}\right) - Kl - D\dot{l} + N\mathcal{N}(t) \tag{1}$$

An explanation of the terms in Equation (1) follows. First, the driving term is $S\tanh(\dot{s}\dot{l})$. Second, there are natural restoring spring (K) and damping (D) forces. Third, there is a noise term $N\mathcal{N}(t)$. The presence of this noise is essential to the operation of the system.

We imagine that the sharpness function $s(l)$ will peak at some l_0. The approximate form of this function, as computed by our chip, will be derived later; for now, we take this sharpness function to be a Gaussian of width σ, peaked around l_0:

$$s(l) = e^{-\Delta^2/2}, \qquad \Delta = \frac{l - l_0}{\sigma} \tag{2}$$

We see that Δ is the displacement from the correct focal point, in units of σ. The constant σ is the depth of field in this model system. Using this $s(l)$ in the dynamical equation (1) we obtain

$$M\ddot{l} = S\tanh\left(\frac{-\dot{l}^2}{\sigma}\Delta e^{-\Delta^2/2}\right) - Kl - D\dot{l} + N\mathcal{N}(t) \tag{3}$$

This equation represents a simple harmonic oscillator with the addition of noise and a novel forcing term. The noise term is essential, because in the absence of noise and in the presence of damping and restoring forces, the system will eventually settle down to $l = 0$, no matter what the form of the sharpness function $s(l)$.

The difference between our model and that of Kotulak and Schor (1986) is that these researchers use $|\dot{s}/\dot{l}|$ as the error magnitude. Their choice is sensible because it compensates, in a sense, for large $\dot{s}$ signals produced by rapid focus changes and not by positional focus errors. We use $\tanh(\dot{s}\dot{l})$ as our error signal because it is difficult to build a well-behaved four-quadrant analog divider. At

† A dot over a quantity indicates differentiation with respect to time.

the time this circuit was built, a product seemed more biologically plausible than did a quotient. Any error function $E(\dot{s}, \dot{l})$ that is positive in the first and third quadrants and negative in the second and fourth quadrants, will retain the correct sign-of-error properties.

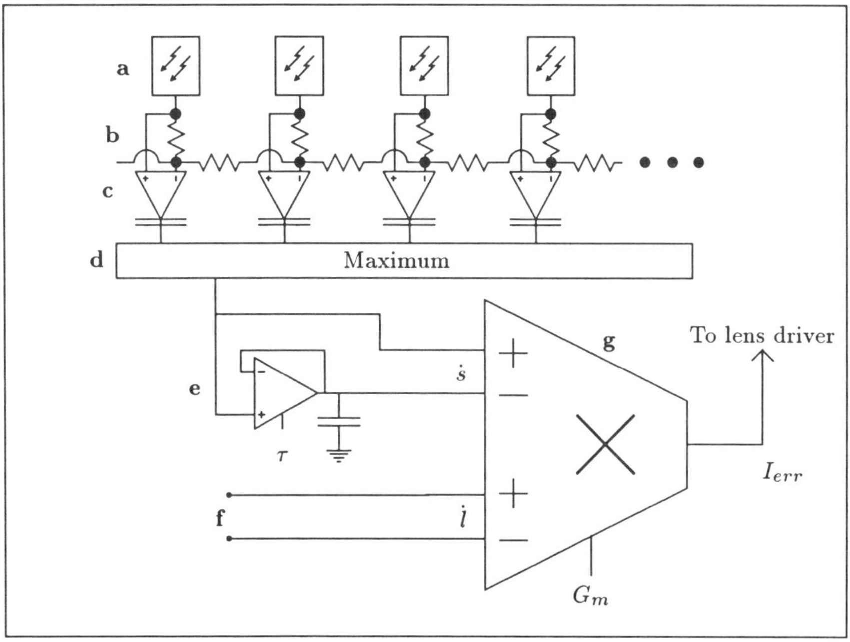

Figure 3. A schematic illustration of the circuitry of the chip.

THE CHIP CIRCUITRY

The chip consists of a one-dimensional silicon retina (Figure 3a,b), an image sharpness computation (3c,d), a differentiator (3e), and an analog multiplier (Figure 3g). All the individual circuits used in this chip have been described in detail elsewhere, so the description here will be confined to a brief functional discussion.

The Sharpness Computation

An image falls on the one-dimensional silicon retina (Figure **3a,b**) (Mahowald and Mead, 1988). Each output of the retina is a differential voltage between the output of a logarithmic photoreceptor (**a**) and the spatial average computed with a resistive network (**b**). We call this differential voltage

V_i for the i_{th} pixel. Each differential voltage is turned into a current by an absolute-value transconductance amplifier (**c**) (Mead, 1989). The i_{th} current is $I_i = I_b \tanh(|V_i/2|)$, where the units of voltage are $kT/q\kappa$. The body-effect factor κ is typically about 0.7. The bias current I_b, and hence the transconductance $G = \frac{I_b}{2kT/q\kappa}$, is set with an externally variable control (Mead, 1989). These I_i are fed into a circuit that computes a voltage s that is logarithmic in the maximum I_i (**d**). This circuit is an adaptation of the winner-take-all circuit (Lazzaro *et al.* 1989), in which the common inhibitory wire encodes the logarithm of the maximum input current. If several I_i are equal and are larger than all other I_j, s will be proportional to the logarithm of the sum of these I_i. This voltage, s, is used as our measure of the image sharpness.

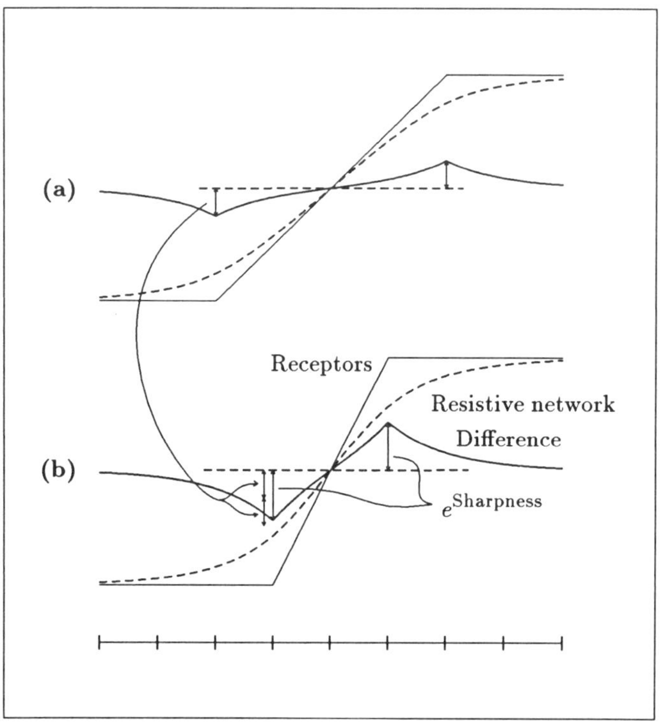

Figure 4. Graphical illustration of the sharpness computation. (**a**) Soft edge. (**b**) Sharp edge. The space constant λ of smoothing is the same for (**a**) and (**b**). When the edge is twice as sharp the maximum difference between the receptors and the resistive net is nearly twice as large, the deficit being caused by the fact that the extent over which the edge is smeared approaches the scale of smoothing.

The sharpness computation is illustrated graphically in Figure 4. The resistive network computes a smoothed version of the log intensities. The space constant λ of smoothing in the resistive network is controllable. The sharpness s is the maximum difference between the local log intensity and the local spatial average computed by the resistive network. This maximum will occur at locations where the slope of the intensity profile changes. When the slope of the intensity profile changes, a distance of $O(\lambda)$ along the resistive network is required for the network to assume the new slope. The equation governing the behavior of a continuous one-dimensional resistive network is

$$\lambda^2 \frac{d}{dx}\left(\frac{dV}{dx}\right) = V(x) - I(x)$$

where $V(x)$ is the voltage on the network at location x, and $I(x)$ is the input voltage, in our case the log intensity (Mead, 1989). In order to change $\frac{dV}{dx}$ by some amount $\Delta(\text{slope})$, the difference $V - I$, integrated over a distance of $O(\lambda)$, must satisfy $\int_{O(\lambda)} (V - I)dx \sim \Delta(\text{slope})$. When the space constant is not too large compared with the extent of the blurred edge, the sharpness s will satisfy

$$s = \log\bigl(\max |V(x) - I(x)|\bigr) \sim \log\left(\frac{\Delta(\text{slope})}{\lambda}\right) \tag{4}$$

When λ is large compared with the extent of the edge, the reported sharpness will not depend on the edge sharpness. On the other hand, when λ is comparable to the receptor spacing, the differences $V(x) - I(x)$ will be small. Circuit offsets will more easily dominate the image induced signals, and will cause the sharpness output to assume a constant value close to the point of optimum focus. The optimum space constant was determined experimentally to be a few receptor spacings.

Since $\Delta(\text{slope})$ is inversely proportional to the distance of the lens from the focal plane (Figure 6), the slope of the sharpness function will be independent of the lens aperture or other geometrical parameters of the system. When the image becomes blurred to the point where image induced signals are comparable in size to circuit offsets, the sharpness signal will flatten out.

Future versions of this chip will probably compute the image sharpness measure by either simply summing the outputs of the absolute value amplifiers, or by computing the maximum first difference in the log intensities.

Time Domain Circuitry

A follower-integrator (Figure 3(**e**)), with externally controllable time constant τ, and transfer function $H(s) = 1/(\tau s + 1)$, computes a delayed version $\tilde{s}$ of the sharpness signal s; the difference $(s - \tilde{s})$ is a good approximation to the derivative $\dot{s}$ for frequencies below $1/\tau$ (Mead, 1989). An external sensor (**f**) gives the lens velocity as a differential voltage $\dot{l}$.

Error Signal Computation

The product of the differential voltage $\dot{s}$ and the velocity signal $\dot{l}$ is computed by a wide-range Gilbert multiplier (g) (Mead, 1989) to produce the error-signal current $I_{err} = S \tanh(\dot{l}) \tanh(\dot{s})$. The multiplier bias current is once again externally controllable, and corresponds to the constant S in Equation (1). This function has characteristics very similar to the function $\tanh(\dot{l}\dot{s})$ used in Equations (1) and (3). The primary difference is that $\tanh(\dot{l}\dot{s})$ saturates more quickly as one moves away from the $\dot{l}$ and $\dot{s}$ axes, away from the origin. The current I_{err} is amplified externally, and is used to drive a solenoid attached to the lens. Finally, the mass of the optical arrangement acts to integrate this error signal with respect to time.

The current version of the chip consists of a 40-pixel array. It was fabricated through the MOSIS foundry in 2μ p-well technology. Each pixel is $165\mu m$ wide.

We tested the function of the sharpness sensor by focusing the image of an edge onto the chip, and varying the distance from the chip to the focal plane of the lens (Figure 5). The output peaked around the point of sharpest focus and fell off on either side, as expected. The width of the peak was consistent with geometrical calculations, as will be discussed later. The slope of the sharpness function was not affected by decreasing the aperture, since the slope of the logarithm of any linear function is identical.

Some theoretical characteristics of of the sharpness output s can be derived as follows. An image that is not in the focal plane of the lens can be represented as the original image convolved with a pill-box shaped kernal. The diameter of this pill-box is the diameter of the circle of confusion at the image plane (Horn, 1968, Horn and Sjoberg, 1981). This procedure uses only geometrical optics; on the scales with which we are concerned, diffractive effects are negligible. Figure 6(a) defines the geometrical parameters. A perfect step edge will be smeared out into an intensity distribution I (Figure 6(b)) given by

$$I = \frac{\pi}{2} - \sin^{-1}(u) - u\sqrt{1 - u^2} \tag{5}$$

where $u = x/r$ represents the distance along the image plane away from the axis of the lens, in units of the radius of the circle of confusion (Figure 6(a)). (The units of intensity here are arbitrary.) The diameter of the circle of confusion is

$$r = \frac{\delta l A}{f} \tag{6}$$

where $\delta l = l - l_0$ is the distance of the image plane away from the focal plane, A is the diameter of the lens aperture, and f is the focal length of the lens.

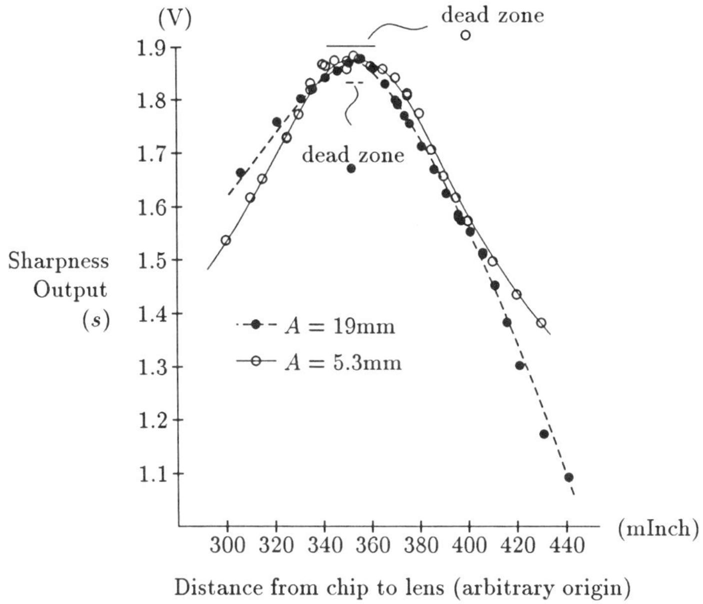

Figure 5. Output from sharpness sensor. The focusing target was a high contrast black and white edge. Using a smaller aperture resulted in a peak that was only slightly broadened, because the space constant λ was several times the receptor spacing d. The curves were hand fitted. The widths of the dead-zones were computed from the geometrical parameters $A = 19mm$ and $A = 5.3mm$, $f = 19mm$, and $d = 165\mu m$, shown in Figure 6. As the distance between the chip and the focal plane is increased to more than is shown in this figure, the sharpness output flattens out. This flattening is due to circuit offsets dominating image induced signals. For smaller apertures (larger depths of field), the flattening occurs farther from the focal plane.

When $\delta l \leq df/A$, s should take on a constant value, because the extent of the edge spans less than one receptor spacing. We compare this prediction with the measurements of the sharpness output shown in Figure 5.

Because measuring image sharpness is equivalent to some measure of the spectral power at the highest spatial frequencies, we can expect some effects of spatial aliasing. This aliasing will cause spurious changes in the reported sharpness, due only to lateral movement of the image, and not to changes in the focus. A scene that is rich in texture will not suffer these spurious

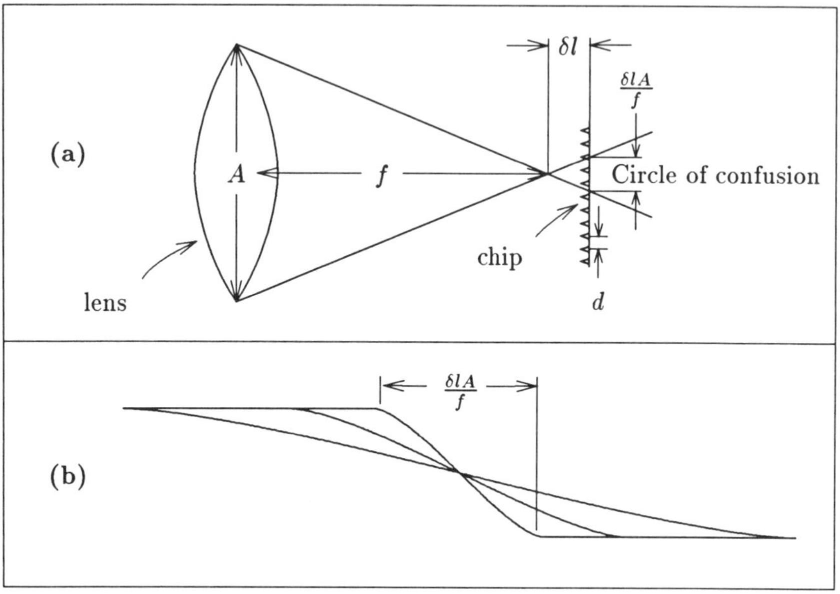

Figure 6. (a) Definitions of dimensions used in the text. The focal length f shown here is the effective focal length for the object being viewed; it is simply the distance from the lens at which the scene is in focus. d is the distance between receptors on the chip. δl is the distance between the chip and the focal plane. (b) Several blurred edges at various distances from the focal plane. The spatial profiles of the intensities are derived in the text.

changes in the reported sharpness; each edge in the image has a different offset relative to the receptor array, and the sharpness sensor chooses the edge with the maximum contrast, as seen by the array. The same principal will apply to a two-dimensional sensor array for a single edge, as long as the edge does not lie along one of the principal axes of the array. An array with randomly jittered pixel locations would be even better, since it has no preferred axes.

The most straightforward elimination of these spurious changes comes from filtering the image before it falls on the sharpness sensor, to eliminate spatial frequency content above the Nyquist frequency for the receptor spacing. This filtering could occur in the human eye, where the optical cut-off frequency has been reported to be matched to the receptor spacing at the center of the fovea (Snyder and Miller, 1977).

Alternatively, we suggest that aliasing will only occur when the scene is in focus. Thus, alias-induced signals can be used as indicators of good focus. In general, the magnitudes of local time and space derivatives of the image, produced by lateral movement of the scene across the sensor, will serve as a good indicator of the focus. By integrating these signals over time and space we can obtain an extremely robust measure of the image quality. This is precisely the type of operation biological retinas can do very well. When the eye is fixating a scene, there are constant slow drift and rapid microsaccadic eye movements (Steinman *et al.* 1973). We suggest that these eye movements may generate signals that are used by the focusing mechanism of the eye.

A PHYSICAL REALIZATION OF THE SYSTEM

A schematic illustration of the system as it is now constructed is shown in Figure 7. The lens actuator is a solenoid with a ferromagnetic plug attached to the lens. The velocity of the lens is sensed with a linear variable transformer. The primary coil is excited with a DC current. The velocity is the differential voltage induced in the secondary coil.

Figure 9, shows records of the lens position obtained from this rather primitive setup using two different-sized apertures. Figure 10 shows the power spectra of these records.

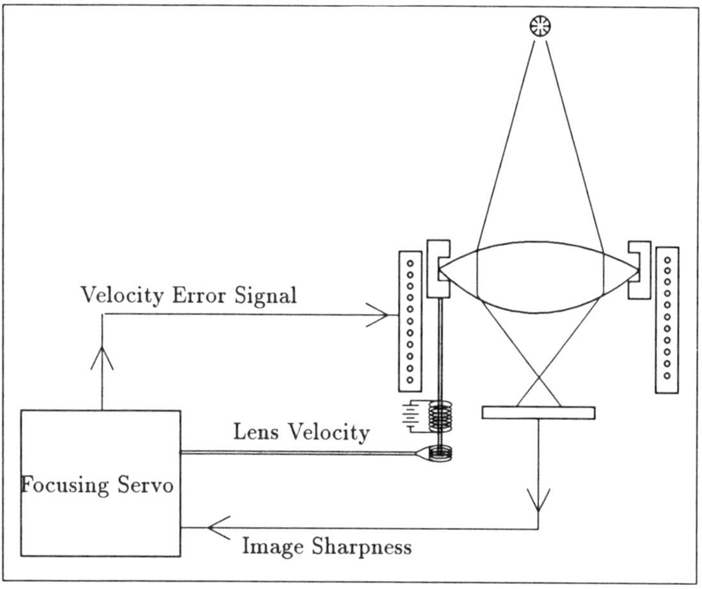

Figure 7. A schematic illustration of the physical interface with the optical system. The output of the chip drives a solenoid which is attached to the lens. The velocity of the lens is sensed with a variable transformer. The mass of the lens serves to integrate, with respect to time, the force signal produced by the chip.

DYNAMICAL PROPERTIES OF THE SYSTEM

Now that we have a dynamical system model we can easily test its explanatory power. Consider Figures 1 and 2; they show the effect a change in depth of field has on human accommodation fluctuations. Figure 8 shows what happens when the model system, represented by Equation (3), is subjected to a simulation of the same change in depth of field. The position l_0 of the focus target is shown by the thin solid line. Halfway through the simulation, the target jumps from one side of the zero point to the other. The zero on the vertical axis represents the equilibrium point in the absence of any focusing target. The only difference between the two simulations is in the width σ of the sharpness function, shown on the left. The results are tantalizingly similar to the records shown in Figure 1 for the human accommodation system. However, the behavior of the dynamical system represented by Equation (1) is dependent in a complex way on the values of the parameters. In the absence of the image-dependent term ($S = 0$), Equation (1) reduces to a simple harmonic oscillator driven by a stochastic noise process. Adding back in the sharpness term ($S \neq 0$) and selecting the correct set of parameters produces the behavior shown in Figure 8, but a different choice of parameters could have led to qualitatively different behavior.

We have distinguished three qualitatively different regimes of operation of our dynamical system. The first represents the behavior shown in Figures 1 and 2 for the human accommodation system and in Figure 8 for the simulations. In this regime, increasing the depth of field decreases the amplitude and coherence of the fluctuations in accommodation. In the second regime, not shown in this report, increasing the depth of field does not substantially change the character of the fluctuations. In the third regime, increasing the depth of field increases the amplitude and coherence of the oscillations. The third regime is shown in Figures 9 and 10 for data taken from our physical realization of the system.

To understand these effects, we make a simple analysis of the relative effects of the sharpness (S), damping (D), and depth of field (σ) parameters in the dynamical system represented by Equation (3), letting the spring constant (K) and noise (N) terms be small. The effect of the damping is to limit the saturated velocity of the lens. This velocity is attained when the nonlinear sharpness term, $S\tanh(\dot{s}l)$ is saturated and $\ddot{l} = 0$. In this state, $|\dot{l}| = S/D$. This condition will be self consistent when the sharpness term is saturated, which will, to first order, happen when

$$\dot{s}l = \dot{l}^2 \frac{ds}{dl} \geq 1. \tag{7}$$

The sharpness function is steepest just outside the dead zone that is caused by finite receptor spacing and other optical blurring. There, $\delta l = df/A$, and

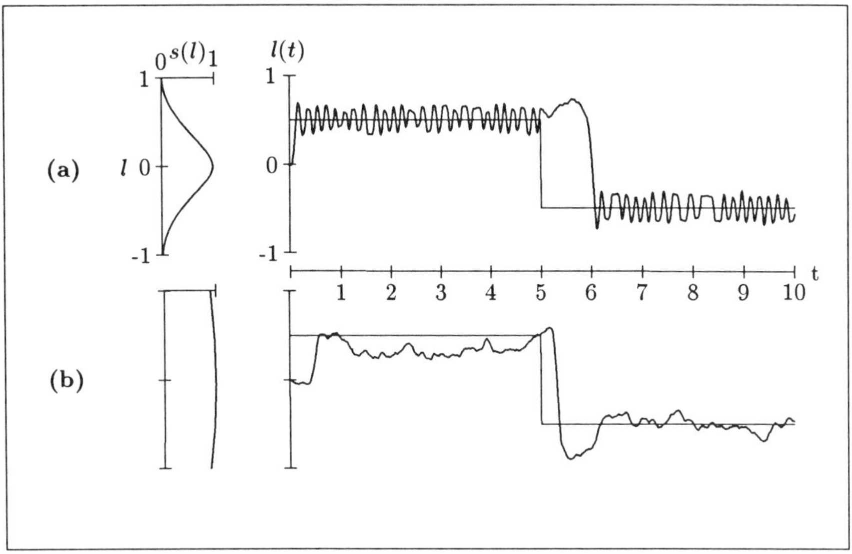

Figure 8. Results of simulations of the dynamical equation (3). The point of optimum focus l_0 switched from $+1/2$ to $-1/2$ halfway through the record. The noise source $\mathcal{N}(t)$ was a Gaussian process of variance 1. The values of the constants were $M = 2.5$, $S = 500$, $K = 1$, $D = 50$, $N = 100$. **(a)** $\sigma = 0.35$. **(b)** $\sigma = 2.5$. The form of the sharpness functions for parts **(a)** and **(b)** are shown on the left. The effect of the depth of field on the amplitude and coherence of the waveform is similar to that shown in Figures 1 and 2.

from Equation (4), $\frac{ds}{d(\delta l)}\Big|_{l=df/A} = \frac{A}{df}$. Using this value for $\frac{ds}{dl}$ and the saturated velocity S/D in Equation (7), we find that the sharpness term will be saturated when $\frac{S^2 A}{D^2 fd} \geq 1$.

As long as this inequality holds, the amplitude of the oscillations will not depend on the depth of field. Intuitively, the velocity will be saturated every time the system crosses the zero point, and an excursion will be halted by the effect of the saturated driving term. This situation corresponds with the second regime of operation mentioned previously.

When the inequality in Equation (7) is no longer satisfied (for example, when the aperture A becomes small enough), we obtain the first regime of operation, in which an increase in depth of field causes the fluctuations to lose their amplitude and coherence. Intuitively, the limiting velocity will no longer saturate the driving term. This leaves the system more susceptible to the built-in noise. This condition corresponds to the behavior shown by the human

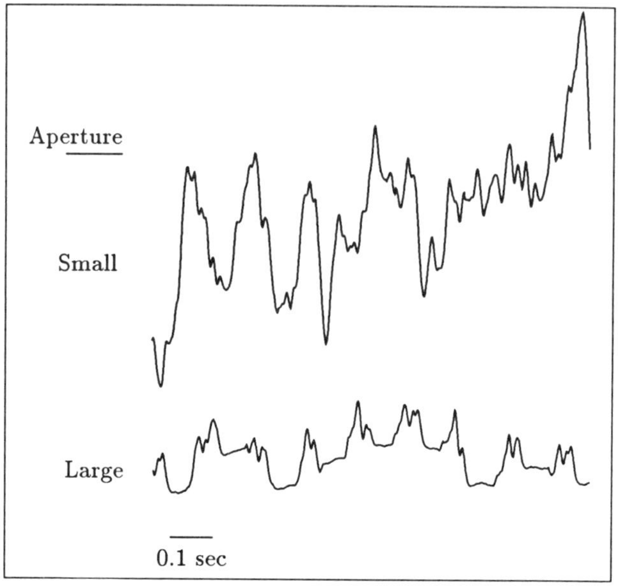

Figure 9. Records of lens motion obtained by integrating the velocity signal numerically.

accommodation system in Figures 1 and 2, and by the simulation records in Figure 8.

The third regime of operation appears when a saturated excursion past the zero point ends because the sharpness derivative $\frac{ds}{dl}$ becomes small, and not because $\dot{l}^2$ gets small (Equation (7)). Figures 9 and 10 show records of the lens motion obtained from our physical realization of the system. We can see from these figures that decreasing the lens aperture increases the amplitude of the fluctuations, opposite to the behavior shown by human accommodation and to the simulation results. In our system, because the sharpness is encoded logarithmically, it is not the slope of the sharpness function that depends on the aperture, but rather the point of defocus where the sharpness function flattens out.

If the model is an accurate representation of the behavior of human accommodation, then we may conclude two principles of that system. First, the human accommodation system probably does not encode the sharpness logarithmically, as we do in our model system. The effect of a change of depth of field in the human system appears as a change in the slope $\frac{ds}{dl}$. Second, human accommodation is optimized so that the strength of the sharpness term is as small as it can be and still allow the system to operate under long depth of

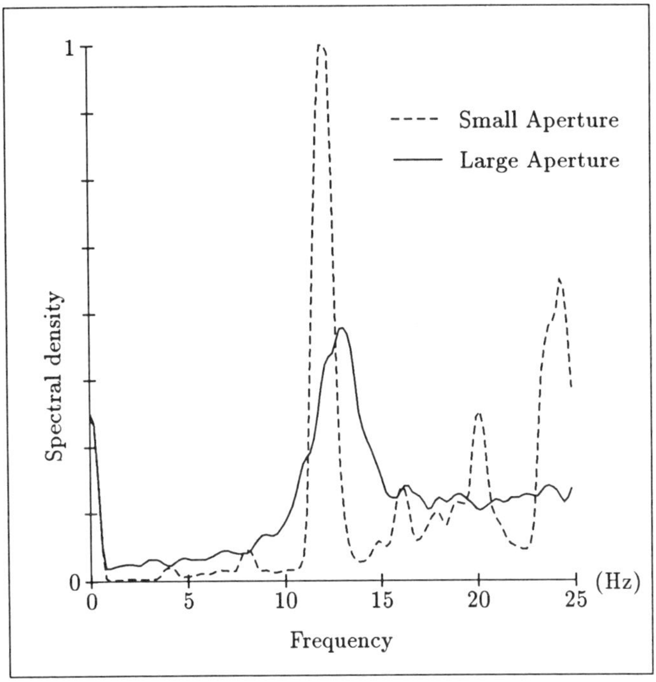

Figure 10. Power spectra for the records in Figure 9. The behavior is opposite to that shown by the human accommodation system in Figure 2. The reason for this difference is discussed in the text.

field. If the sharpness term was any larger, the fluctuations would be larger than they need to be under conditions of short depth of field.

DISCUSSION

The human accommodation system provides a simple example of a biological control system in which needed information may be generated actively by the motor system, in the course of control. The system described here represents a control system of a relatively unexplored variety. The system is unstable; small oscillations around the desired state are amplified until a nonlinearity becomes saturated. The system relies on noise for initiation of control. We might hope that these characteristics would circumvent many of the problems of gain control about which designers of control systems worry, but probably these problems are simply pushed into another arena.

We have omitted many features of the human accommodation system. The concept of volition is alien to our formulation; our system has no means of deciding that it would like to alter its focus in a particular direction. In the human accommodation system, volition certainly plays an important role in directing the apparatus toward the desired state. Also, our system has no concept of a linkage between vergence and accommodation. In the human accommodation system, there is strong coupling between vergence eye movements and accommodative response (Johnson *et al.* 1982). Our system's lens is mass dominated, and the damping forces are small relative to the restoring forces. The human lens system is probably spring dominated, with large damping forces (Ejiri *et al.* 1969). The role of the slow reaction time (1/3 sec.) in the human accommodation system has not been worked out, but could signify the presence of a neural integrator like that seen in the vestibulo-ocular reflex (Robinson, 1981). Alternatively, there might not be any neural or physical integration of the error signal; the error signal might affect the velocity of the lens directly. The use of a saturating nonlinearity is biologically plausible (Marg, 1955). The presence of noise is essential for operation of our system; human accommodation is quite noisy even in the absence of any stimulus (Johnson *et al.* 1984). The hypothesis that time domain information is being used is just that – a hypothesis, albeit an attractive one. A related hypothesis is that image sharpness is the primary image-quality cue that the human accommodation system uses. We have used sharpness in our model of accommodation, but this computational convenience does not indicate that other cues, such as chromatic or spherical aberration, might not be used as well.

Several features of the system we have built appear repeatedly in silicon models of the nervous system, and are worth pointing out. Quantities are scaled logarithmically, so that a large dynamic range is compressed into a workable operating range. Nonlinear aspects of operation can be advantageous. Time, as an intrinsic dynamical variable, appears naturally when we use analog computation. The active generation of time domain information may turn out to be useful in other contexts.

Acknowledgments

I thank Misha Mahowald, Carver Mead, John Harris, John Wyatt, and Berthold Horn for critical comments. I thank Hewlett-Packard for computing support, and DARPA and MOSIS for chip fabrication. This work was sponsored by the Office of Naval Research and the System Development Foundation.

References

Campbell, F.W. (1959). The accommodation response of the human eye. *Brit. J. of Physiological Optics.* **16**:188–203.

Campbell, F.W. and Westheimer, G. (1959). Factors influencing accommodation responses of the human eye. *J. Opt. Soc. Amer.* **49**:568–571.

Campbell, F.W., Robson, J.G., and Westheimer, G. (1959). Fluctuations of accommodation under steady viewing conditions. *J. Physiol.* **145**:579–594.

Denieul, P. (1982). Effects of stimulus vergence on mean accommodation response, microfluctuations of accommodation and optical quality of the human eye. *Vision Res.* **22**:561–569.

Ejiri, M., Thompson, H.E., and O'Niell, W.D. (1969). Dynamic viscoelastic properties of the lens. *Vision Res.* **9**:233–244.

Fujii, K., Kondo, K., and Kasai, T. (1970). An analysis of the human accommodation system. *Technology Reports of Osaka University.* **20**:221–236.

Helmholtz, H.v. (1924). *Treatise on Physiological Optics,* Vol. 1. Menasha: Optical Society of America, p. 191.

Horn, B. (1968). Project MAC: Focusing. *MIT Artificial Intelligence Memo.* No. 160.

Horn, B. and Sjoberg, R.W. (1981). The application of linear systems analysis to image processing. Some notes. *MIT Artificial Intelligence Memo.* No. 100.

Johnson, C.A., Post, R.B., and Tsuetaki, T.K. (1984). Short-term variability in the resting focus of accommodation. *Opthal. Physiol. Opt.* **4**:319–325.

Kotulak, J.C. and Schor, C.M. (1986). A computational model of the error detector of human visual accommodation. *Biol. Cybernetics.* **54**:189–194.

Kotulak, J.C. and Schor, C.M. (1986b). The accommodative response to subthreshold blur and to perceptual fading during the Troxler phenomenon. *Perception.* **15**:7–15.

Lazzaro, J., Ryckebusch S., Mahowald, M.A., and Mead, C.A. (1989). Winner-Take-All circuits of $O(n)$ complexity. In Touretsky, D.S. (ed), *Advances in Neural Information Processing Systems* 1. San Mateo, CA: Morgan Kaufman, pp. 703–711.

Mahowald, M. and Mead, C.A. (1988). A silicon model of early visual processing. *Neural Networks.* **1**:91–97.

Marg, E., Reeves, J.L. (1955). *J. Pot. Soc. Am.* **45**:926 (Fig. 1).

Mead, C.A. (1989). *Analog VLSI and Neural Systems.* Reading, MA: Addison-Wesley.

Robinson, D.A. (1981). The use of control systems analysis in the neurophysiology of eye movements. *Ann. Rev. Neurosci.* **4**:463–503.

Snyder, A.W. and Miller, W.H. (1977). Photoreceptor diameter and spacing for highest resolving power. *J. Opt. Soc. Am.* **67**:697–698.

Steinman, R.M., Haddad, G.M., Skavenski, A.A., and Wyman, D. (1973). Miniature eye movements. *Science.* **181**:810–819.

Weale, R.A. (1960). *The Eye and Its Function.* London: Hatton.

8

A FOVEATED RETINA-LIKE SENSOR USING CCD TECHNOLOGY

J. Van der Spiegel, G. Kreider
Univ. of Pennsylvania, Dept. of Electrical Engineering
Philadelphia, PA 19104-6390

C. Claeys, I. Debusschere
IMEC, Leuven, Belgium

G. Sandini
University of Genova, DIST, Genova, Italy

P. Dario, F. Fantini
Scuola Superiore S. Anna, Pisa, Italy

P. Bellutti, G. Soncini
IRST, Trento, Italy

ABSTRACT

A CCD imager whose sampling structure is loosely modeled after the biological visual system is described. Its architecture and advantages over conventional cameras for pattern recognition are discussed. The sensor has embedded in its structure a logarithmic transformation that makes it size and rotation invariant. Simulations on real images using the actual sensor geometry have been performed to study the sensor performance for 2D pattern recognition and object tracking.

A CCD imager consisting of 30 concentric circles and 64 sensors per circle, whose pixel size increases linearly with eccentricity has been fabricated. The central part has a constant resolution with 102 photocells. The CCD is made in a three phase buried channel technology with triple poly and double metal layers. Preliminary results of the testing are given showing the validity of the design.

INTRODUCTION

Charge coupled devices are structures that can move charge packets or perform simple mathematical operations on the packets. They consist of an array of MOS capacitors which, when pulsed with a high voltage, drive the silicon in deep depletion. The charge packets stored near the semiconductor-insulator interface of the MOS capacitors are the samples of an analog signal. A proper sequence of pulses on the gates of the capacitors will shift the packet from one capacitor to the next, resulting in a delay, summation, subtraction, multiplication or filtering of the charge packets. These properties, together with the photosensitivity of silicon material, make CCD technology well suited to implement a variety of structures and functions. Indeed, since their invention twenty years ago, CCD's have been widely used as image sensors [1-2], memories [3], filters [4-5] and analog signal processors [6-7] as well as retinal neural net processors [8]. CCD technology offers the capabilities of high density, high speed and low power implementation of these functions.

The area of solid-state imagers, dominated by CCD cameras, has been mainly driven by color video cameras and machine vision for automatic inspection. Megapixels CCD's are currently available and cameras for high definition color television (HDTV) are being developed [9].

In this paper we will describe the features, design and implementation of a foveated retina-like sensor realized in a CCD technology. The proposed architecture is very different from the conventional image sensors and is intended for use in object recognition, pattern classification and tracking. Other image sensors and artificial retinas have been built previously. The pioneering work by C. Mead has resulted in neural based vision systems which incorporate several features of the biological visual system [10-12]. These imagers have been fabricated with CMOS technology. The sensor described in this paper complements these developments in two aspects: it makes use of an alternate technology, i.e. CCD technology, and captures the sampling structure of the biological visual system. Although no other biological features have been incorporated in the current design, the technology and architecture lend themselves well to include charge coupled neural processing elements.

RETINA-LIKE SENSOR

Imaging has been done traditionally with sensor arrays that have a uniform resolution over its entire photosensitive area. This is typical for systems used to generate an undistorted image such as for TV or video. However, this method is not necessarily the optimum strategy from a viewpoint of efficient visual perception and recognition. In situations where one needs real-time coordination between sensory perception and motor control, such as machine vision for robotics, target recognition or autonomous navigation, one is not interested in an exact reproduction. Quick detection, localization and tracking of an object is of prime importance in the first place. Zooming-in on the right object, once it is detected and tracked, will provide the required fine details. This calls for a non-conventional sensing scheme optimized to perform scene analysis rather than reproduction.

It is instructive in this respect to look at the biological visual system. Research on the anatomy of the eye has revealed that the photoreceptors are not uniformly distributed over the retina. The cone density shows a peak in the center of the visual field and decreases towards the periphery. The receptive field, receiving projections form the foveal area, of a simple cell in the visual cortex (area V1) receives inputs from 12 by 20 receptors in the fovea. These cells can resolve linewidths at least equal to the receptor spacing. In order to be able to obtain this high resolution, an area of equally densily spaced receptors is required. However as one moves away from the center the inverse magnification (defined as the inverse of the cortical surface devoted to a given portion of visual space) and the field size increase as a function of retinal eccentricity [13-15]. It has been suggested that the mapping of the retinal surface into the striate cortex can, under certain conditions, be described by a complex logarithmic function [16]. This may provide the scale and rotation invariances observed in the biological visual system.

The sampling structure of the retina-like imager is loosely modeled after the early stages of the biological visual system to capture the logarithmic mapping discussed above. Rather than using a uniform square grid, as is done in commercial cameras, the retina sensor has a highly non-uniform sampling grid. The center, called the fovea, has a constant resolution while the peripheral sensors are organized in a circular fashion whose size increases linearly with eccentricity [17]. A schematic representation of the periphery is shown in Fig. 1a.

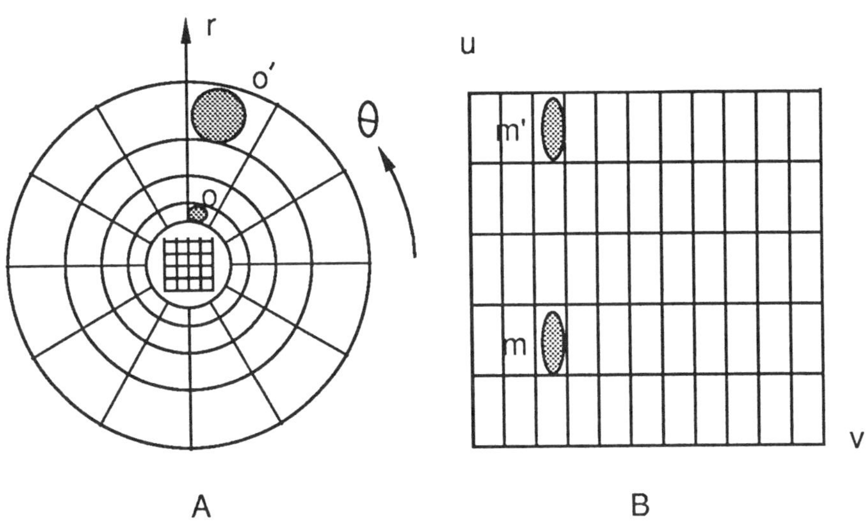

Figure 1 (a): Schematic of the retina-like sensor. The middle part consists of a constant resolution photosensitive area (fovea) while the peripheral area is organized in a circular fashion whose sensor size increases linearly with eccentricity; (b) Cortical representation of the image as a result of the retino-cortical logarithmic mapping of the peripheral sensors.

A point in the retinal plane can be described by its polar coordinates (r,θ):

$$z = r \exp(j\theta).$$

After mapping the retinal plane into a Cartesian plane (cortical) as shown in Fig. 1b, the new coordinates become,

$$\begin{aligned} w = \ln(z) &= \ln(r) + j\theta \\ &= u + jv \end{aligned}$$

This complex logarithmic transformation has interesting properties for pattern recognition. Its advantages for 2D shape recognition [18,19] and motion stereo [20] have been pointed out earlier. Its main characteristics are size and rotation invariances. These properties are characteristic of the human visual system as well: an object does not change its perceived shape when it is rotated or scaled. Rotation of the image impinging on the sensor will result in a linear translation along the v axis in the Cartesian plane. Similarly, an enlarged image will be represented in the cortical plane as a translated version of the original image. This can be easily seen as follows: if one scales the object by a factor "a" such that

$$z' = ar \exp(j\theta)$$

the transformed image becomes,

$$w' = u + jv + \ln(a).$$

This is schematically shown in Fig. 2 for two circles of different size. The scale invariance is also valid for any object if its translation is scaled in the same proportion as the object. This is a direct consequence of the logarithmic mapping. This can be easily proven using the graphical representation of Fig. 1, where objects o and its scaled version o' are mapped into features m and m'.

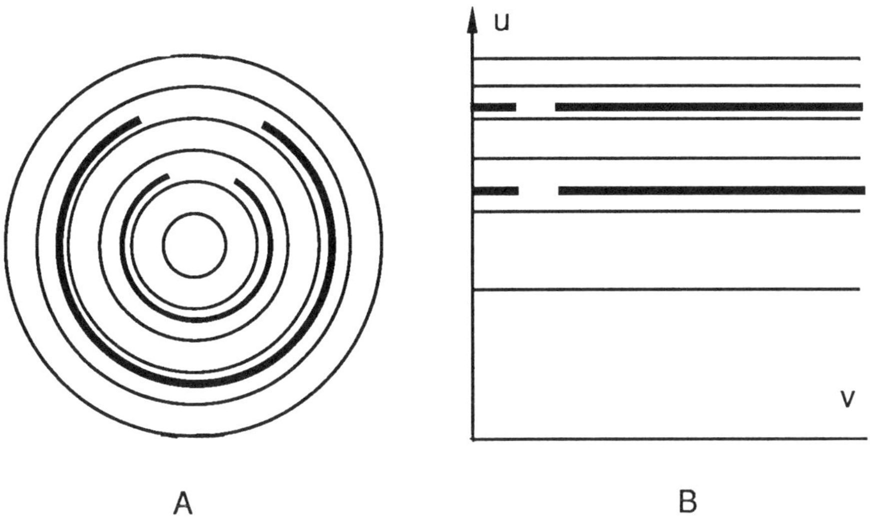

Figure 2: Logarithmic mapping of two circles of different size (a), illustrating that magnification results in a simple translation in the cortical plane (b).

The mapping of the sampling structure is similar to a Mellin transform whose modulus is independent of scale changes [21]. :

$$f(r) \;\text{---->}\; M(\omega) = \int_0^\infty f(r) r^{-j\omega-1}\, dr$$

$$f(ar) \;\text{---->}\; a^{-j\omega} M(\omega)$$

This transformation can be obtained by taking the Fourier transform of the scaled function in which the coordinate r is replaced by,

$$r = e^u \quad \text{or} \quad u = \ln(r), \;\text{and}\; dr = e^u\, du$$

$$M(\omega) = \int_{-\infty}^{\infty} f(e^u) r^{-j\omega u}\, du$$

Because of the scale and translation invariant properties of the Mellin and Fourier transform, respectively, both transforms have attracted considerable attention for image correlation in optical systems [22]. Also the optical Mellin transform in conjunction with a circular photodiode array has been used for scale invariant pattern classification [23]. The imaging system, described in this paper, has the logarithmic transformation, required for the size invariance, built into the hardware structure of the sensor and does not require any additional computation.

It has been shown that the algorithm results in a leading edge invariant representation and in intensity invariance [24]. Also in the use of camera motion, the apparent motion of the objects caused by a translation along the optical axis will result in a translation along the θ axis of the transformed image. However, the invariances are true only under certain conditions: the scaling should occur along the optic axis or rotation around the optic axis. It should be pointed out that the transform is not translation invariant, a characteristic also observed by the human visual system: indeed as one moves his eyes away from an object it looses its exact shape until it gets completely blurred.

Besides its scale and rotation invariances the retina-like sensor has also the important property that it reduces the amount of information considerably.This is important because one of the main limitations in pattern recognition is the immense amount of data to be processed. Reducing the data as much as possible and as early as possible is a high priority. Because the transformation is built into the sensor's hardware there are no computations required to transform the incoming images. The sensor's sampling structure acts as a spatially inhomogeneous filter that attenuates the importance of points away from the origin. In this respect it can be considered as a structure that performs a modified Fourier transform in which an exponential weighting function has been included into the transform [16, 19].

Simulations of the retina-like imager have shown that this representation facilitates scene analysis in comparison to a conventional constant resolution sensor, in particular when objects of widely different sizes have to be recognized [18]. Research in the field of active vision and sensory-motor coordination have exploited different strategies. It is important to be able to look at a scene with a wide visual field and at the same time to sample certain areas in greater detail. The peripheral vision provides the wide view and is typically used for alerting and guidance purposes. For this task one wants to work with a minimum amount of data in order to be able to respond quickly. Once the object is tracked, one likes to explore it with a high resolution. The fovea performs this function and corresponds to the focus of attention. The sensor, as shown in Fig. 1, combines both functions in one and thus provides a good compromise between high resolution and data reduction. This approach allows us to reduce the amount of data at the very early stage, i.e. at the sensor level. Further reduction is possible by incorporating into the same structure some primitive functions such as motion detection and edge detection. The architecture with the large peripheral area lends itself well for such modifications. A second version of the retina-like sensor that has these functions built in is under development.

DESCRIPTION OF THE CCD RETINA-LIKE SENSOR

Overall Sensor Organization

The sensor is functionally divided into two main areas as schematically shown in Fig. 1a. The middle section is a constant resolution CCD that is built with minimum feature sizes. It provides the high resolution fovea. Around the fovea is the peripheral area whose function is to provide a wide field of view while keeping the amount of data to a minimum. It realizes also the logarithmic mapping required for the scale invariance. The cell dimensions increase with eccentricity according to,

$$h_i = a_o \cdot s^{k-1} , \quad k=1,2,3, ..., m$$

where a_o is the minumum cell dimension, k is the concentric circle number, counting from outward, s is a scaling factor, and m is the number of concentric circles. In our design, s is made equal to 1.094. The size of the cells increases then from 30 for the smallest to 412 micrometers for the outer circle. The sizes and the corresponding cell density as a function of eccentricity are shown in Fig. 3 and Fig. 4, respectively, illustrating the logarithmic nature of the sampling structure.

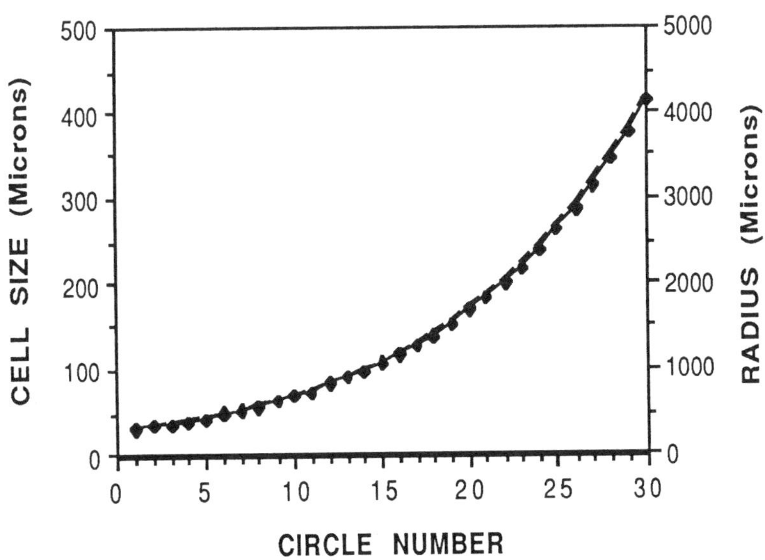

Figure 3: Size of the photosensitive cells as a function of the circle number.

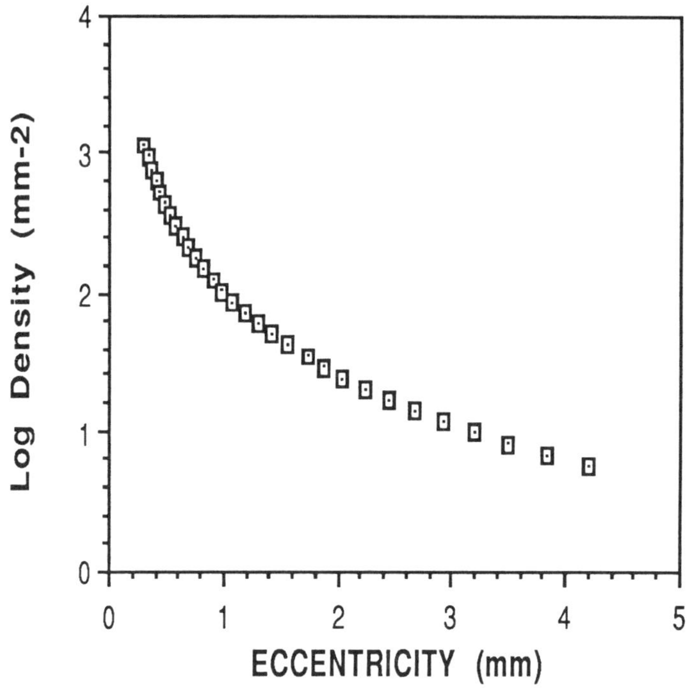

Figure 4: Cell density versus eccentricity

Peripheral Imaging Area

The peripheral imaging section consists of thirty concentric circles whose radii increase exponentially, according to Fig. 3. These circles are grouped in three arrays, each consisting of 10 circles clocked differently. There are 64 sensors per circle for a total of 1920 pixels. The pixel density is highest in the innermost rows with a pitch of 30 μm. The scaling of the pixel size is quite dramatic as can be seen in Fig. 3 with a size increasing from 30 to 412 μm going from the inner to the outer circles. Figure 5 gives a schematic picture of the actual geometry implemented in the imager. Notice the three blocks of concentric circles. The central part is the fovea. The slice cut out of the periphery to the right is required to provide the read-out and coupler structures, as described later.

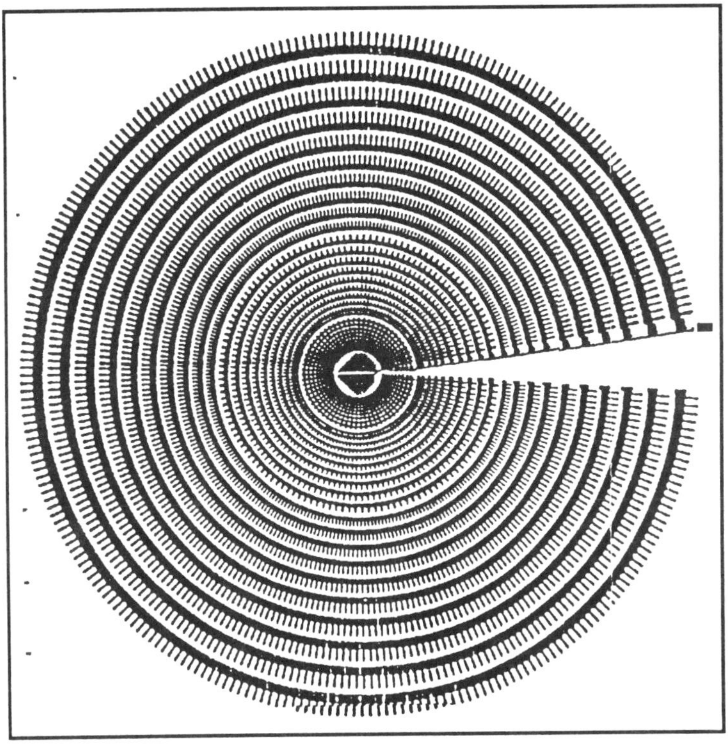

Figure 5: Schematic of the actual peripheral and foveal areas. The ratio between the smallest and largest radius is 13.7; diameter is 0.94cm

The imager has a modified interline transfer structure. The basic cell consists of a photoreceptor, a transfer gate separating the sensor from the CCD channel, and a three phase CCD shift register. A diode is used as a photoelement rather than a semi-transparent CCD because of its better photosensitivity [25]. The cell dimensions are 30 by 30 μm. An array of these cells forms the basic pattern for all circular CCD's. These arrays, scaled and rotated by computer, are easily pieced together. The photosensitive area is defined by an aluminum light shield allowing to make the photosites any desired shape. A more detailed description of the actual implementation of the peripheral structure is given in [26].

Fovea

The inner part of the imaging structure consists of a relatively high density square array that is tilted 45 degrees. A unique layout, read out and clocking scheme allows routing of the clock lines into this highly populated central area to read out the charge packets effectively without taking too much real estate [26]. The sensors are stacked, forming a checker board pattern. Fig. 6 gives a schematic view of the fovea's photosites pattern. The total number of sensors in the fovea is 102. The effective pixel pitch is 60 μm.

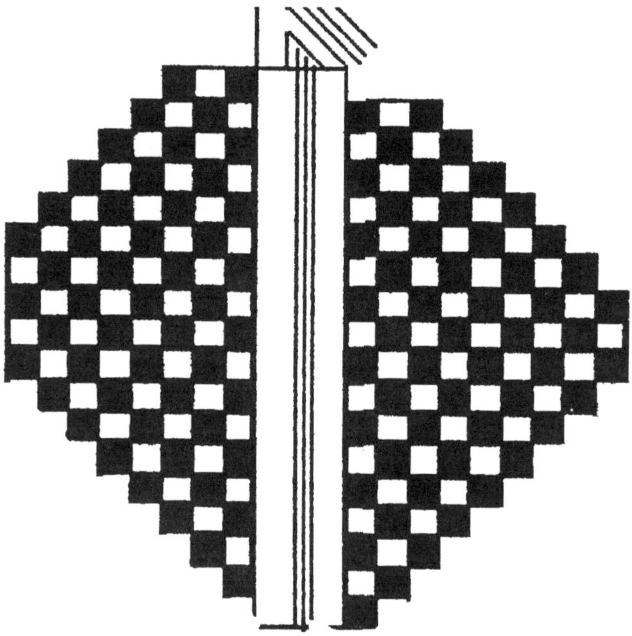

Figure 6: Schematic drawing of the fovea, illustrating the sensor geometry. The total number of pixels is 102.

Read-out Structures

Two blocks are required to read out the data. One is a linear CCD register called the radial CCD (RCCD). It has a cell size of 30 μm and is a three-phase buried channel CCD. This register is used to read out the circular CCD's. This eliminates separate outputs for each circular CCD thus reducing the amount of output pins at the expense of lower read out speed. Another version with parallel read out of the circular CCD's under design will provide a faster readout, to be used for tracking fast moving objects.

The charge packets of one cell of each circular CCD are transferred simultanously after being scaled into the RCCD that is read out quickly before the next charge packets from the circular registers are transferred into it. However, because the radial CCD has a constant cell size, while the cells of the circular CCD increase exponentially one needs to scale the incoming charge packets before reading them in the RCCD. This is done by using a fixed-ratio divider between the radial and circular register. It accounts for the scaling of the sensor, preserving information but reducing the size of the charge packets. The RCCD and coupling cells cut out a part of the peripheral area (Fig. 5). This slice occupies an area of two and a half photosites.

The output of the RCCD consists of a diode that can be reversed biased by a reset transistor. When charge packets are dumped into the output node the diode will discharge, resulting in a voltage drop that is detected by a source follower. The maximum voltage drop for a full charge packet is about 1 V. The output consists of a double sourcefollower and is designed for speed and low noise.

The total chip dimensions are 11mm by 11mm. It is mounted in a 30 pin package with optical transparent window. A photograph of the whole sensor is shown in Fig. 7. The chip is fabricated in IMEC's triple poly, three phase, buried channel CCD techology. Two poly layers are used for the three phases in the shiftregisters, while the third layer defines the transfer gates.

Figure 7: Microphotograph of the retina-like sensor, including a central fovea region. The chip area is 11x11 mm^2.

A detailed view of the fovea is shown in Fig. 8. The stripe in the middle is a read-out register that feeds into the radial CCD. The ten inner circular CCDs can be seen around the fovea.

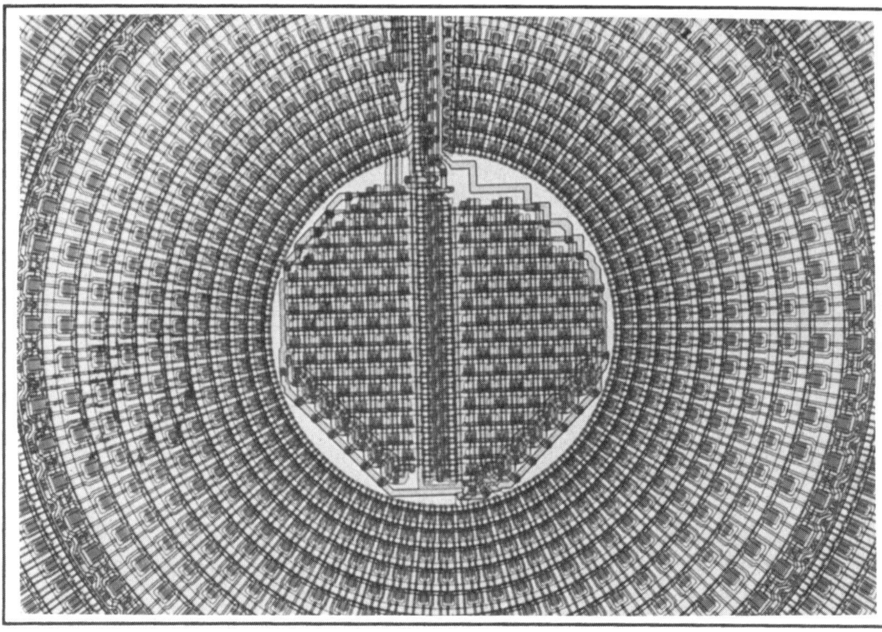

Figure 8: Photograph of the fovea, consisting of 102 photosensitive cells, and the first ten concentric circles.

Driving Electronics

One of the complications of this architecture is the relatively large amount of clocks and control signals to read out and synchronize the charge flow. Up to 18 different clocks are required. When the sensor has to be used as part of a moving platform for tracking purposes, it is important to minimize the number of wires and external interconnections. Also the dimensions and weight of the clock drivers should be small. For this reason an integrated clocking system has been developed that generates all the required clocks. It has been fabricated in a 2 μm CMOS process. The chip is fully custom designed in order to reduce the amount of real estate and power dissipation as much as possible. The total chip area is less than 3 mm^2. A photograph of the chip is given in Fig. 9 [27]. This chip will be mounted together with the CCD imager on a lightweight substrate and incorporated into the motor control platform. The chip is fully functional. Measured outputs of the controller chip is shown in Fig. 10.

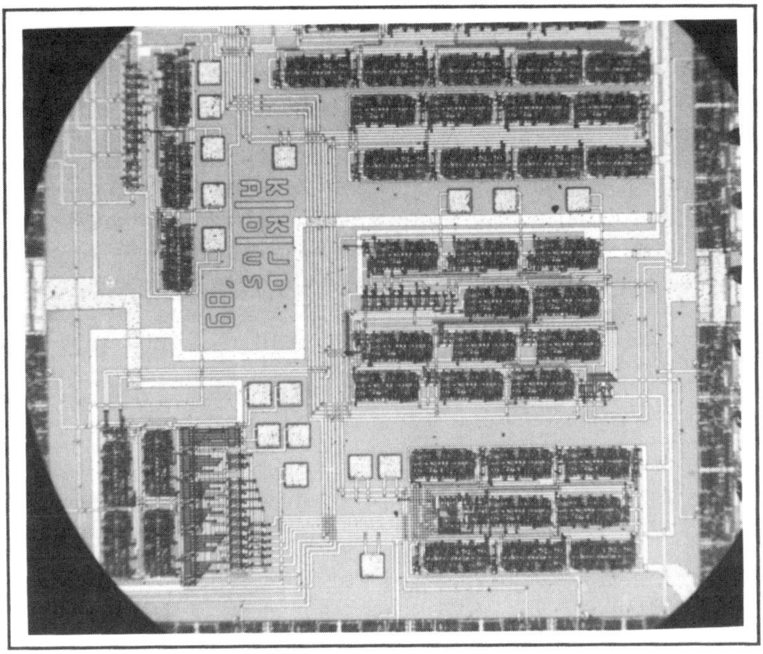

Figure 9: Microphotograph of the controller chip of the retina sensor, generating all required pulses to drive the sensor; the size is $1.7x1.7\ mm^2$.

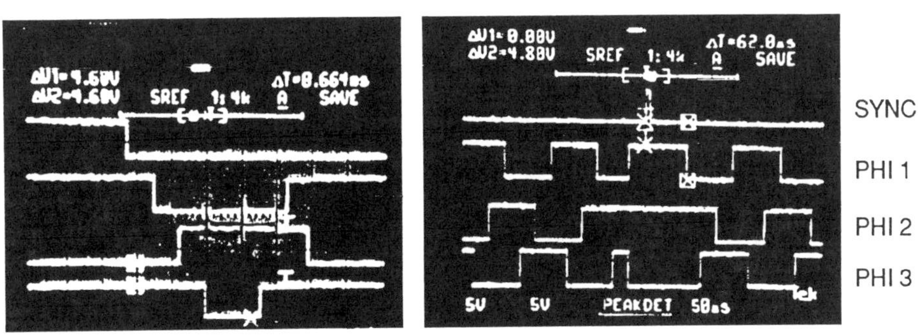

Figure 10: Measured output pulses of the controller chip, showing (a) the pulses required for performing the adding and weighting function ; (b) the sync pulse and the three phase clocks for the middle ten circles at the start of a new frame [27].

SIMULATION RESULTS OF THE RETINA-LIKE SENSOR

The actual sampling structure of the CCD sensor is slightly different from the one proposed in Fig. 1 due to the slice which is cut out of the peripheral area and the actual pixel geometry. In order to evaluate the effect of these differences, several simulations were done using the actual sampling grid and pixel geometry. Fig. 11 shows how an image sampled with the actual CCD (Fig. 11a) is mapped into the cortical plane (11b). The mapping of a magnified image (Fig. 11c) is shown in Fig. 11d. One notices the scale invariance. The jagged line is the result of the finite pixel size. The slight assymetry is caused by the slice cut out of the peripheral area.

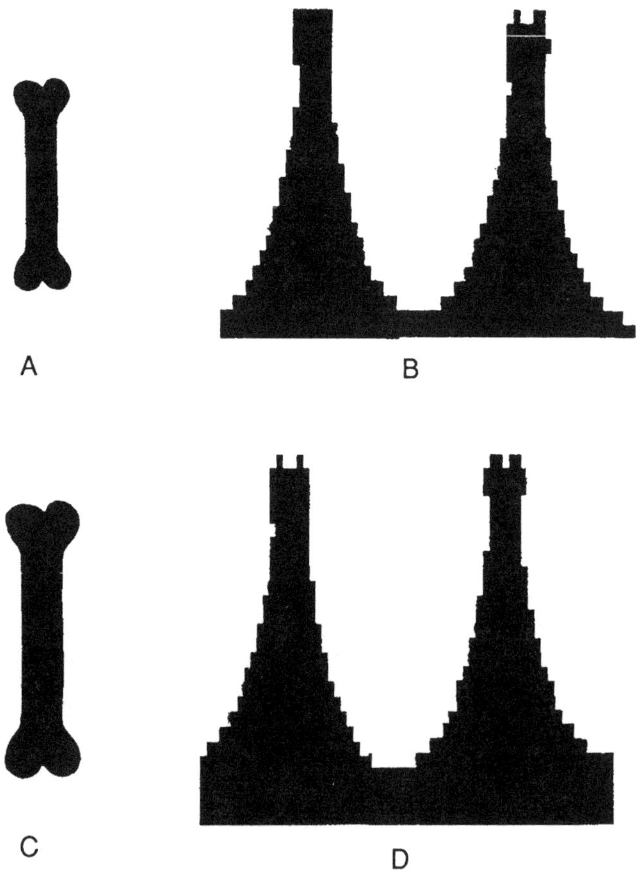

Figure 11: (a) Original image and (b) cortical image; (c) magnified image of (a) and corresponding cortical image (d). Notice the invariances in the cortical image.

In Fig. 12 a more complex image and its scanned representation is shown.

A.

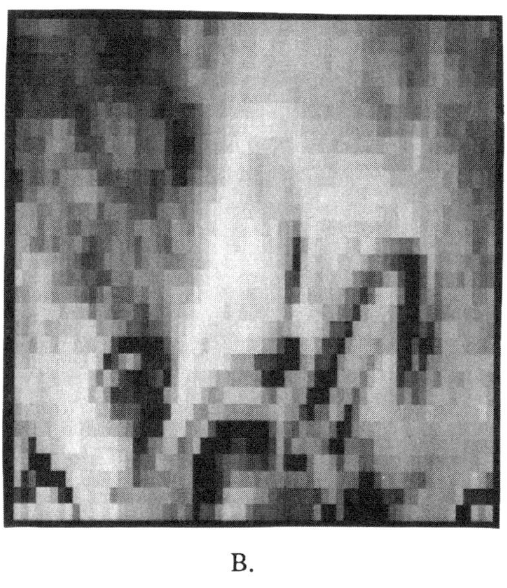

B.

Figure 12: (a) A complex image (the "Apollo of Veio"), (b) and its mapped image on the cortical plane.

A

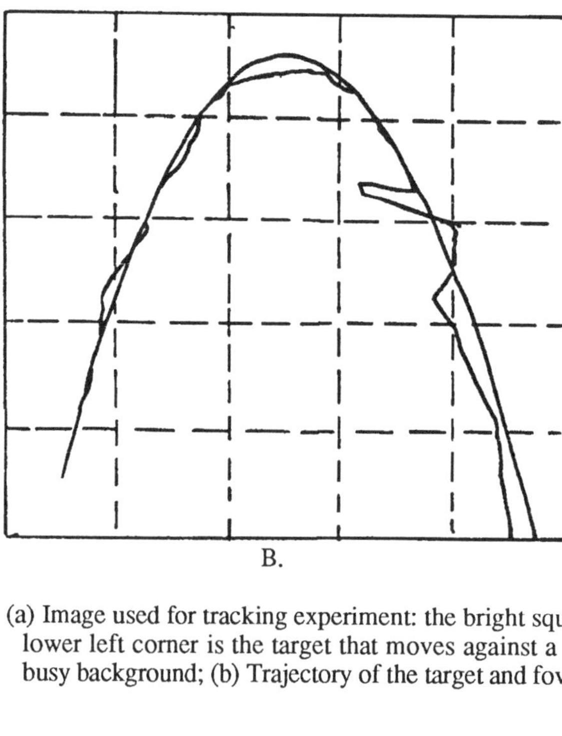

B.

Figure 13: (a) Image used for tracking experiment: the bright square in the lower left corner is the target that moves against a relatively busy background; (b) Trajectory of the target and fovea.

One of the most obvious and immediate applications of the proposed sensor is the tracking of moving targets. The small amount of information and the relatively wide visual field allow the reduction of the execution time of the segmentation algorithms and the computation of the position and speed of the moving object. For this reason anthropomorphic algorithms simulating the ocular movements of a human subject gazing at a target in motion have been developed and tested [28]. The motor strategy proposed is a succession of saccadic and smooth pursuit movements, in analogy with the human behavior: as soon as the target enters the visual field its position and velocity is computed, and the target is "foveated" and the sensor moved at the computed speed. At every new frame the measure of the target speed is updated and used to correct the sensor motion. If the object is too far from the fovea, the measured position is used to drive a correction saccade. Of course, during the tracking, it is necessary to subtract from the optic flow, which is computed from the cortical images, the velocity field of the points belonging to the background; this last field is due to the camera motion and so computable from motor commands. In Fig. 13 an example of the performance is shown. The original image is given in Fig. 13a. In this case the picture of the Apollo of Fig. 12 serves as a background image and the small bright square target on the lower left corner serves as the moving target. The target is moved over a parabolic trajectory at constant speed. The output is shown schematically in Fig. 13b. In this figure the trajectory of the fovea is shown together with the parabolic trajectory of the target. As can be seen the superposition is good apart from a big saccade to the left on the right branch of the parabola (caused by a wrong estimate of the optical flow) which is corrected soon after. We can compare this experiment with one that makes use of a constant resolution camera. The size of the background image comprizes about 512x512 pixels and the target 70x70 pixels. The number of photosites in the retina camera is 1920. If these sensors would have been distributed evenly in a constant resolution camera it would correspond to an array of about 45x45 cells. This is less than the size of the target. One could project the whole image (background) on the 45x45 array, but in that case the target would get lost in the background, making it impossible to track it. Thus, the advantage of the retina-like sensor is that it can cover a large field without sacrificing the resolution of the target. This make the tracking faster and more efficient than with a constant resolution camera.

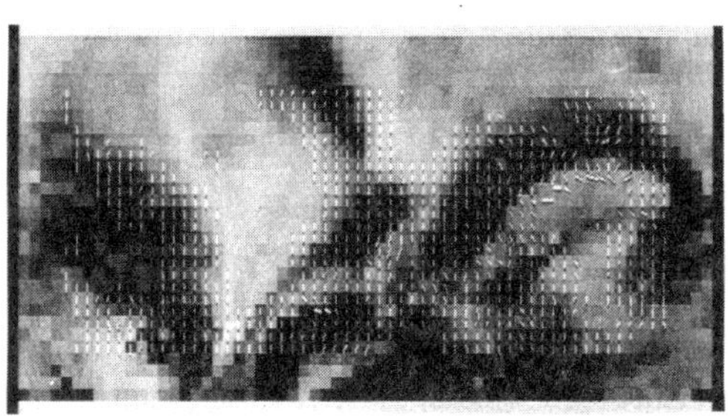

Figure 14: Optical flow computed from a sequence of 9 images

In Fig. 14 the result of a different experiment is presented. The picture shows the optical flow computed from a sequence of 9 images acquired during a motion of a camera along the optical axis and toward the picture of the "Apollo" is shown. It can be noticed that the optical flow vectors are parallel as is expected.

MEASUREMENTS

The devices were fabricated on 125 cm p-type Si wafers with 56 sensors per wafer. Measured technological parameters are typical of the IMEC process. Short circuit testing gives a high yield. Preliminary functional testing demonstrates that the sensor works correctly and well. Proper clocking is the most important factor in operating the chip - controlling the substrate voltage, the high and low voltages and rise and fall times of the clocks can critically affect performance.

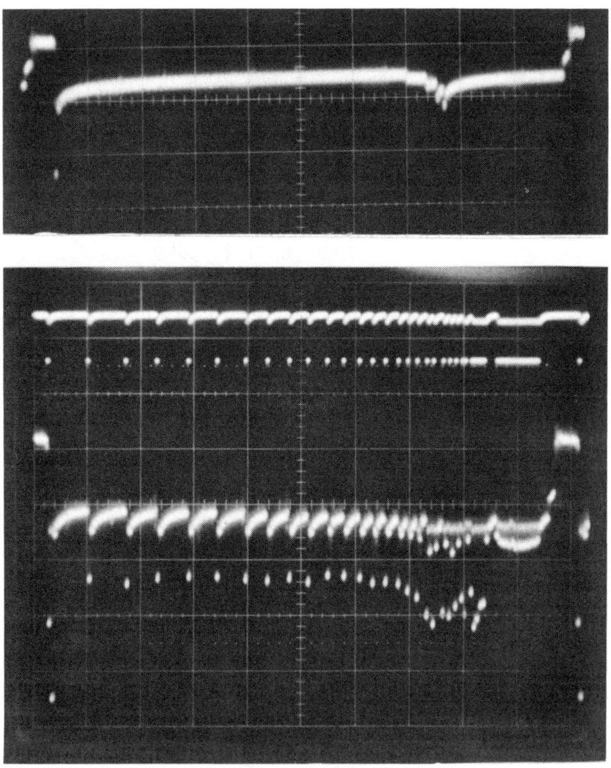

Figure 15: (a) Measured output waveforms corresponding to one column of the CCD; (a) dark output and (b) the output under uniform illumination.

Figure 15 is an unfiltered output signal found from the chip. Figure 15a is the dark response which is uniform and low. A couple of cells around one quarter from the right of the trace have a larger output than the others. It is a systematic signal that occurs in each sample and which is caused by the layout. It can be easily corrected for in software. The layout of the next version which is currently being fabricated was modified to eliminate this problem. Figure 15b is the sensor output under partial uniform illumination. The top trace is the signal valid, indicating that a charge packet has been placed on the output. The bottom trace is the actual output. Several features need to be mentioned. The row (circle) 11 peak is still present, which is due to the same layout effect as was observed in the dark current. At row 30 (shown at the beginning of the trace) an improperly generated clock pulse causes a large fixed output signal. This can be corrected by changing the clock driver circuit. Rows 12-29 show a uniform response. The inner section, rows 1-10 gives different outputs. This is also the result of the layout where the size of the photosensor is scaled differently in comparison to the coupler ratio. This had to be done due to lack of space. The displayed pattern corresponds to the actual design and is the same for each tested sensor. Also this will be modified in the next version. Finally, the response at the right side of the photo is the fovea. A good uniformity is obtained but the output signal is considerably smaller. This is due to a different photosensitive area.

Photo 16a and 16b are the dynamic response of the sensor to a projected light spot. The outputs of five columns are shown on one trace. The single output pulse in the 3rd block is the response to the light spot. There is very little crosstalk between neighboring circles. However, one can see a small pulse in the different columns which indicates that there is some charge overflow between the neighboring diodes or cells on the same circle. This is a function of the bias voltages and light intensity. No special effort was made to reduce this effect during these preliminary measurements. In Fig.15d the pulse has shifted to the left corresponding to the light spot which has been projected in the next column.

Measurements of the output signal versus light intensity gives a linear response. The spectral response peaks around 580 nm. These preliminary results are promising and are demonstrating that the basic concept of the structure works. The results have been used to make an optimized redesign of the sensor. The sensitivity at 400 and 700nm is about 75% of the peak value.

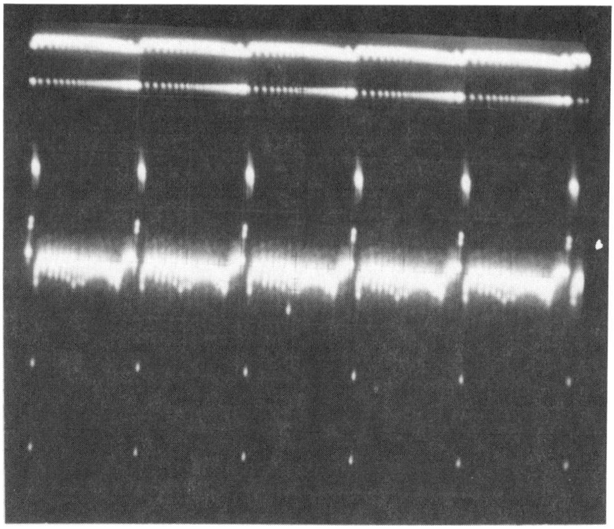

A.

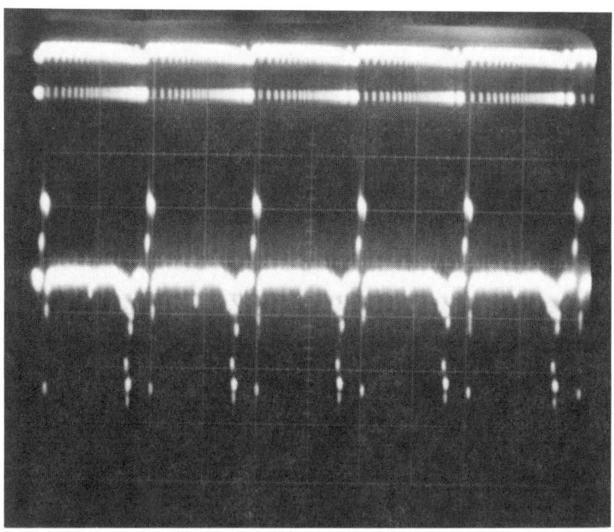

B.

Figure 16: The output of five columns with one pixel illuminated in column three (a) and in column two (b). The top traces are the signal valid outputs.

DISCUSSION AND CONCLUSION

A space-variant CCD sensor which capture certain aspects of the sampling structure of the human visual system has been discussed and presented. The imager consisting of a fovea and peripheral areas has been designed and fabricated. The preliminary results are promising and show that the basic elements of the imager are functioning properly. The detailed characterization is going on and will be used for further optimization of a second prototype.

The "anatomy" of the sensor provides pseudo scale and rotation invariances. In this respect it is similar to a Mellin transform. Simulations have been performed using the actual imager's sampling grid. It has been shown that this new structure offers advantages for scene analysis and tracking purposes over conventional cameras. It is the purpose to incorporate some primitive feature extraction function on the next chip. The retina-like imager will then serve as input to an analog neural network that is independenlty being developed [29]. Also the question of image decomposition in its primitives and how to implement this in the neural network is being investigated [30].

The sensor will also be useful as part of an optical image processing system for pattern recognition or correlations.

Acknowledgements

The authors like to thank Prof. G. Declerck and R. Van Overstraeten for their generous support for the fabrication of the CCD at IMEC (Leuven) facilities. They also are thankful to Prof. P. Mueller for the discussions on the biological visual system. The financial assistance of the NSF, CNR, the NATO and the Penn-Leuven Exchange is acknowledged. The authors thank also those who helped with the fabrication of the sensor and driving electronics.

References

1. M. G. Collet, "Solid-State Image Sensors", Sensors and Actuators, **10**, 287-302 (1986).
2. H. Shiraki, "Recent Progress of CCD Image Sensors", Proc. 6th Sensor Symp.,153-159 (1986).
3. H. Veendrick , F. Steenhof, G. Davids, P. Hartog, E. Holle, K. Lismore, B. Pham, K. van der Sanden, A. Slob. J. Slotboom, G. Streutker, H. Van der Veen, W. Wiertsema, and A. van Zanten, "An 835 Kbit Video Serial Memory", Tech. Digest 1988 Intern. Solid State Circuits Conf.,44-45, (1988).
4. J. Tieman, T. Vogelsong and A. J. Steckl, "Charge Domain Recursive Filters", IEEE J. Solid-State Circuits, **SC-17**, 597-605 (1982).
5. Y. Maki, T. Kondo, A. Izumi, I. Matsuda, T. Fukuda, T. Narabu.," A CMOS-CCD Comb Filter with Dropout Compensation for a VCR", Tech. Digest 1988 Intern. Solid State Circuits Conf., 46-47 (1988).

6. A. Chiang, P. Bennett, B. Kosicki, R. Mountain, G. Lincoln, J. Reinhold, " A 100ns CCD 16-point Cosine Transform Processor", Tech. Digest 1987 Int. Solid State Circuits Conf., 306- 307, (1987).
7. E. Fossum, "Charge Domain Analog Signal Processing for Detector Arrays", Nucl. Instr. and Methods, **A275**, 530-535 (1989).
8. A. Chiang, "A CCD Retina and Neural Net Processor", Workshop on "Hardware Implementation of Neural Nets and Synapses", eds. P. Mueller, C. Mead, Lau and J. Hopfield, San Diego,171-182, (1988).
9. T. Nobusada, M. Azuma, H. Toyada, T. Kuroda, K. Horii, K.Otsuki, G. Kano, "Frame Transfer CCDD Sensor for HDTV Camera", Tech. Digest 1989 Int. Solid State Circuits Conf., 88-89 (1989).
10. J. Tanner and C. Mead, "A correlated Optical Motion Detector", Proc. Conf. Advanced Res. in VLSI, MIT, Cambridge, Jan. 1984, P. Penfield (ed), Artech House, Dedham, MA, p. 57 (1984).
11. M. Sivilotti, M. Mahowald, C. Mead, "Real Time Visual Computations using Analog CMOS Processing Arrays", Stanford Conf. on VLSI, P. Losleben (ed), MIT Press, Cambridge, p. 295 (1987).
12. C. Mead, "Analog VLSI and Neural Systems", Addison-Wesley Publ., Reading, MA (1989).
13. D. Hubel and R. Wiesel, "Receptive Fields and Functional Architecture of the Monkey Striate Cortex", J. Physiol., **195**, 215-143 (1968).
14. B. Dow, A. Snyder, R. Vautin, R. Bauer, "Magnification Factor and Receptive Field Size on Foveal Striate Cortex of the Monkey", Experimental Brain Research, **44**, 213-228, (1981).
15. V. H. Perry and A. Cowey, "The Ganglion Cell and Cone Distributions in the Mokey's Retina: Implications for Central Magnification Factors", Vision Res., **25**, 1795-1810 (1985).
16. E. Schwartz, "Computational Anatomy and Functionall Architecture of Striate Cortex: A Spatial Mapping Approach to Perceptual Coding", Vision Res., **20**, 645-669 (1980).
17. G. Sandini, V. Tagliasco " An Anthropomorphic Retina-like Structure for Scene Analysis", Comp. Graphics and Image Proc.,**14**, 365-372 (1980).
18. L. Massone, G. Sandini, V. Tagliasco, "Form-Invariant Topological Mapping Strategy for 2D Shape Recognition", Comp. Graphics and Image Proc.,**30**,1169-188 (1985).
19. C. R. Carlson, R. W. Klopfenstein, C. H. Anderson, "Spatially Inhomogeneous Scaled Transforms for Vision and Pattern Recongition", Optics Letters, **6**, 386-388 (1981)
20. R. Jain, "Motion Stereo using Ego-motion Complex Logarithmic Mapping", Rep.no. RSD-TR-3-86, Center for Res. Integr. Manuf., Univ. Michigan, Ann Abor, 1986.
21. M. McDonnell, "A Clarification on the Use of the Mellin Transform in Optical Pattern Recongition", Opt. Commun.**25**, 320-322 (1978).
22. D. Casasent, D. Psaltis, "Position, Rotation, and Scale Invariant Optical Correlation", Appl. Optics, **15**, 1795-1799 (1976).
23. T. Yatagai, K. Choji, H. Saito, "Pattern Classification using Optical Mellin Transform and Circular Photodiode Array", Opt. Commun, **38**, 162-165 (1981).

24. R. Messner and H. Szu, "An Image Processing Architecture for Real Time Generation of Scale and Rotation Invariant Patterns", Comp. Vision, Graphics and Image Proc., **31**, 50-66 (1985).
25. J. Van der Spiegel, J. Sevenhans, A. Theuwissen, J. Bosiers, I. Debusschere, G. Declerck, "Study of Different Sensors for High Resolution Linear CCD Imagers", Sensors and Actuators, **6**, 51-64 (1984).
26. I. Debusschere , E. Bronckaers, C. Claeys, G. Kreider, J. Van der Spiegel, P. Bellutti, G. Soncini, P. Dario, F. Fantini, G. Sandini, "A 2D Retinal CCD Sensor for fast 2D Shape Recognition and Tracking", 5th Int. Solid-State Sensor and Transducers Conf., Montreux, June 25-30, 1989; to be publisched in Sensors and Actuators, **20** (1989).
27. K. Aricanli and K. Desai, "Controller Chip for the Retinal Sensor", Report EE442, Dept. Electr. Eng., Univ. Pennsylvania, Philadelphia, Apr. (1989).
28. G. Sandini, F. Bosero, F. Bottino, A. Ceccherini, "The Use of an Anthropomorphic Visual Sensor for Motion Estimation and Object Tracking," Proc. of the OSA 1989 Topical Meeting on Image Understanding and Machine Vision, June 12-14 (1989).
29. P. Mueller, J. Van der Spiegel, D. Blackman, T. Chiu, T. Clare, J. Dao, C. Donham, T.-P. Hsieh, M. Loinaz, "A Programmable Analog Neural Computer and Simulator ", in Advances in Neural Information Processing Systems 1, D. Touretzky (ed) , Morgan Kaufmann Publ., San Mateo, CA, 712-719 (1989).
30. P. Mueller, D. Blackman, P. Spiro, R. Furman, " Neural Computation of Visual Images", Proc. IEEE 1st Annual Int. Conf. Neural Networks, **4**, 75-87, San Diego, June 1987.

9

Cooperative Stereo Matching Using Static and Dynamic Image Features

M.A. Mahowald and T. Delbrück
Department of Computer Science
California Institute of Technology
Pasadena, California, 91125
e-mail: stereo@hobiecat.caltech.edu

Visual experience is intrinsically subjective. The manifest unity of our perceptions belies the indeterminacy of our sensations. A single pattern of excitation on the retinal receptors is consistent with many possible worlds of objects. The simple problem of determining object brightness exemplifies the ambiguities inherent in the retinal image. The photons incident on a given receptor may indicate the presence of a bright or dark object depending on the overall level of illumination. In mediating automatic gain control, the horizontal cells of the retina express an assumption about the nature of the illuminant. These neurons in the most peripheral part of the visual system take the first step toward interpretation of the image. By a process of lateral inhibition, they discount the effect of illumination and determine the brightness of an object relative to that of nearby objects. At all levels of complexity, the visual system interprets each data point within the context of the scene. This interpretation is consistent with the input data and with the internal structure of the system. Through evolution, the structure of the visual system provides correspondence between the mental image and the objective world.

As a result of the work of many engineers and scientists, artificial visual systems are also evolving. System synthesis can lead to a better understanding of natural systems, since it demands a concrete formulation of the relationships among representation, system architecture, and the visual world. Marr, who pioneered a computational approach to vision, has been a major influence in this field (Marr, 1982). He viewed vision as a form of information processing that must be based on constraints that come from consideration of the visual world. He described a process as taking place within a definite representational framework according to a specific algorithm.

In 1976, Marr and Poggio proposed a collective algorithm for stereopsis that could perform feature matching between two images (Marr, 1976). The crux of the problem is that images with dense, homogeneous textures contain many identical features. Each feature can be matched to several features in the corresponding image, but only one match is correct. It is not possible to find the correct match without considering several features simultaneously. Constraints derived from an analysis of space, objects, and projective geometry specify interactions among features that permit identification of a correct match. These constraints are based on assumptions about the nature of the visual world.

Marr's algorithm describes a locally connected, fedback, nonlinear system suitable for realization in an electronic medium. The system architecture embodies the assumptions of the algorithm. Using this architecture, we have synthesized an analog CMOS circuit that finds regions of correspondence between two one-dimensional images in real time. Experiments on simplified images demonstrate the circuit's tolerance to transistor mismatch resulting from the collective nature of the algorithm.

We have fabricated and tested two chips that use this correspondence circuit. One of these chips performs stereo matching based on static contrast features in the image. The other chip takes advantage of natural time constants of the analog system, and uses time derivatives of intensity to drive the correspondence circuitry.

Evolving electronic artificial vision increases our awareness of the role of the physical nature of the system. Algorithms for vision must be consistent not only with the external world, but also with the properties of the computational medium.

STEREOPSIS

The review presented here will be brief. A more complete description of the problem of stereopsis can be found in (Julesz, 1971; Poggio and Poggio, 1984).

Binocular vision generates two images of a scene, one from each eye. Because the two eyes regard the scene from different points of view, they may differ in their impression of the relationships between objects. Figure 1 shows two eyes of an observer in cross-section. The lens of the eye projects an image of targets in three-dimensional space onto the the surface of the retina. The position of a feature in the projected image depends on the visual direction of the target. Visual direction is the angle of the line of sight between the retina and the target. When the eyes fixate on a point, the locus of points in space having equal visual direction in both eyes is called the *horopter*. All targets on the horopter are said to have zero disparity, because they project to corresponding points on the two retinas. Targets closer to the viewer than the horopter have crossed (negative) disparity, and targets more distant than the horopter have uncrossed (positive) disparity. The principle task of stereopsis is to determine the disparity between corresponding regions of the images from the two eyes. This information, along with the state of vergence of the eyes and the eyes' separation in the head, specifies the distance between the target and the viewer.

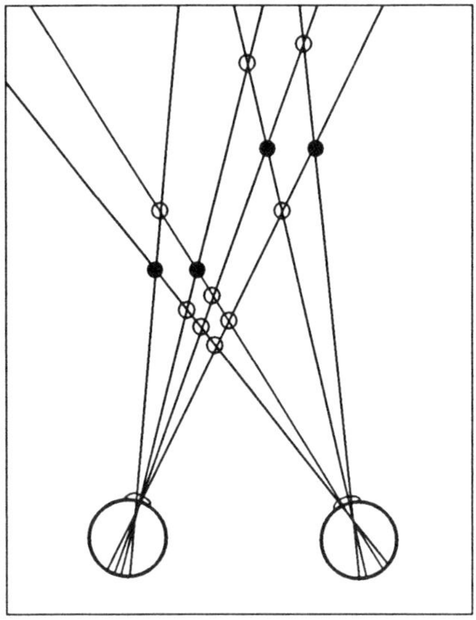

Figure 1 Stereopsis. This figure illustrates the projection of an image of four identical targets (dark disks) onto the right and left eyes of an observer. The lines going through the lens connecting each target with the retina are lines of sight. The intersections of the lines of sight indicate possible target positions in space. False targets (transparent disks) are located at the intersections of lines of sight that originate from different targets in the two eyes.

Calculation of disparity requires the determination of correspondence between features on the two retinas. This task would be straightforward if features could be identified uniquely. However, the random-dot stereograms developed by Bela Julesz (Julesz, 1960) demonstrate that the human visual system can compute disparity even when there are many identical features in close proximity (Figure 2). It has been shown that occlusion events are primary cues to the determination of depth (Shimojo *et al.* 1985). Researchers in artificial vision, however, have focused on the matching of stationary features visible to both eyes in attacking the problem of stereopsis (Poggio and Poggio, 1984).

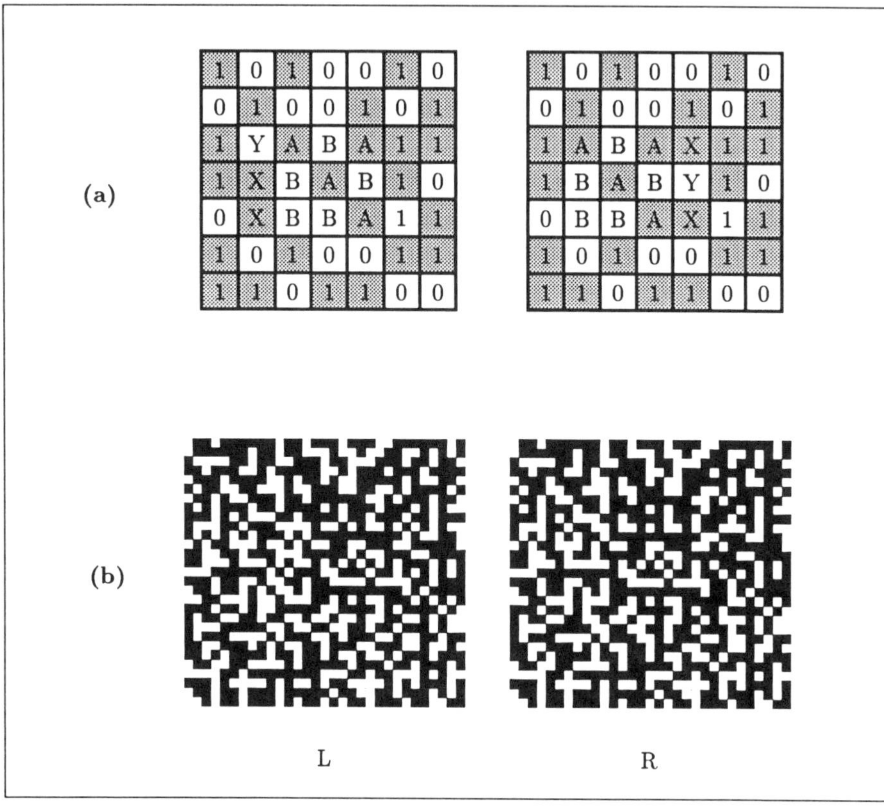

Figure 2 Random-dot stereograms. **(a)** Making a random-dot stereogram. A random pattern of 1s and 0s is generated to be presented to the left eye. An identical copy of the pattern is made for the right eye, except that a central square region within the image (labeled with As and Bs) is displaced to the right. When the two images are fused, this square region will appear closer than the background. Occluded areas (areas having no counterpart in the opposite eye's image) are labeled with X's and Y's. (Modified from (Julesz, 1971)). **(b)** A random-dot stereogram showing a raised square. You can fuse the stereogram by letting your eyes diverge as though you were looking at infinity. Your left eye should see the pattern on the left, and your right eye should see the pattern on the right. The primary difficulty is focusing on the paper while your eyes are diverged. Myopic readers will find it helpful to remove their glasses.

THE MARR/POGGIO STEREO ALGORITHM

Marr and Poggio (Marr, 1976) describe a collective stereo algorithm that succeeds in finding the disparity in random-dot stereograms. Their algorithm is designed to find the disparity in a region of limited depth around the horopter. The algorithm can be used for stereo matching in one- or two-dimensional images. One-dimensional stimuli can be used to test many of the basic features of a stereo matching algorithm, because the eyes are displaced from each other along a line; therefore, only disparities along a line contribute to depth perception. We can understand the algorithm by considering three simple rules derived from the properties of images and of physical surfaces. These rules allow the correct correspondence of features on two retinas to be found in the presence of false matches.

The first rule is that features in the two images must be similar to correspond to each other. For example, dark features within one image should correspond to dark features in the other image. This rule is called the *compatibility* constraint. Psychophysical evidence suggests that the human visual system obeys some form of compatibility constraint. For example, people are unable to fuse images of reversed contrast (Julesz, 1971).

The second rule is that a feature from one image should correspond uniquely to one feature from the other image. This constraint is derived from the fact that a point on a surface has only one spatial location at a given time. The *uniqueness* constraint is violated in the case of transparent surfaces, when an image feature is a combination of points from two physical surfaces.

The third rule is based on the observation that objects, being cohesive, occupy a localized region of depth. The *continuity* constraint used in the algorithm assumes that surfaces of objects are oriented parallel to the viewer so that changes in disparity will be rare, occurring only at surface boundaries.

The representational framework of the algorithm is depicted in Figure 3. Input to the algorithm is provided by two one-dimensional retinas that encode the horizontal positions of features in the image. In the output representation, real space is divided into a grid of discrete positions. The spatial dimensions of distance (disparity) and horizontal position are encoded by a two dimensional array of correlators. Activation of one of these correlators indicates the presence of target in the region of space to which it corresponds.

The constraints of the algorithm are embodied in the connections among elements. The compatibility constraint is implemented by feedforward input to the correlators from the retinas. The correlators receive two inputs, one from each eye. A correlator is stimulated by a nonlinear combination of its inputs; the two inputs must signal the presence of a similar feature in order to drive the correlator. False matches may be formed between features that do not correspond — features that are not actually generated by the same target in space. False matches are suppressed by feedback connections within the correlator array that implement the continuity and uniqueness constraints.

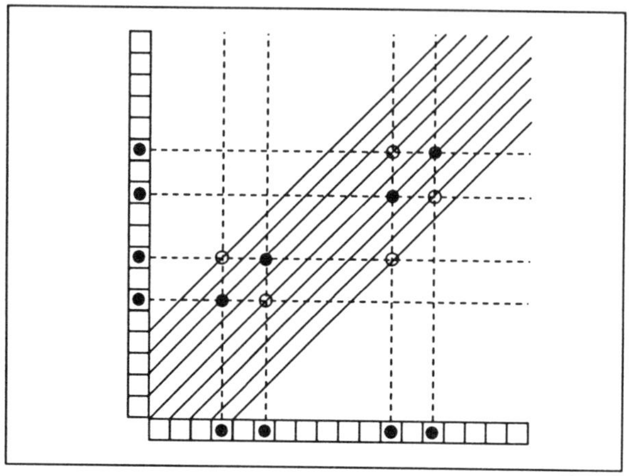

Figure 3 The representational framework of the cooperative algorithm. Depicted is the algorithm's response to the scene illustrated in Figure 1. The left and right retinas are represented on the vertical and horizontal axes. Correlators (not shown explicitly) are arranged in lines that are angled 45 degrees from the axes. Each line corresponds to a single disparity. The outputs of each retina are transmitted along lines of sight running perpendicular to the retinal axis. Each correlator receives input from the two pixels whose lines of sight intersect at its location. Wherever there are similar features from the right and left eye, they form a match (circle) at the intersection of their two lines of sight. Targets at the same distance in space form matches located on the same disparity plane. Open circles are false matches, and filled circles are true matches. Inhibitory interaction among correlation elements run along lines of sight (dotted lines). Solid lines along disparity planes indicate positive coupling between correlators.

Positive coupling between correlators at the same disparity encourages the continuity of the solution. The positive coupling is implemented in the original algorithm by a fixed set of connections among correlators in the same disparity plane. If a correlator is active, it drives the correlators to which it is coupled. Correlators at the correct disparity are all driven by retinal inputs with matching features and so tend to be active. Because correlators at the true disparity are surrounded by correlators that are likely to be active, they receive not only positively correlated retinal inputs but large inputs via positive coupling between correlators. The false matches receive less input from neighboring correlators.

The uniqueness constraint states that each feature from one eye can be matched to only one feature from the other eye. This constraint is implemented by an inhibitory interaction among correlators that receive input from the same pixel. This interaction performs a winner-take-all function that selects true matches by suppressing all but the maximally driven correlator. The final state of activation of the correlator array indicates the positions of genuine targets in space.

THE CHIP

The Marr/Poggio stereo correspondence algorithm is well suited to a circuit implementation because it requires only local connectivity. The system architecture leads to a physical structure that instantiates the representational framework of the algorithm. Figure 4 provides a simplified view of the chip architecture. The chip correlates the outputs of two one-dimensional retinas, and the representation of the solution expands into the second dimension of the silicon surface. The circuits are analog nonlinear elements that compute the correspondence between regions of the retinal images in real time.

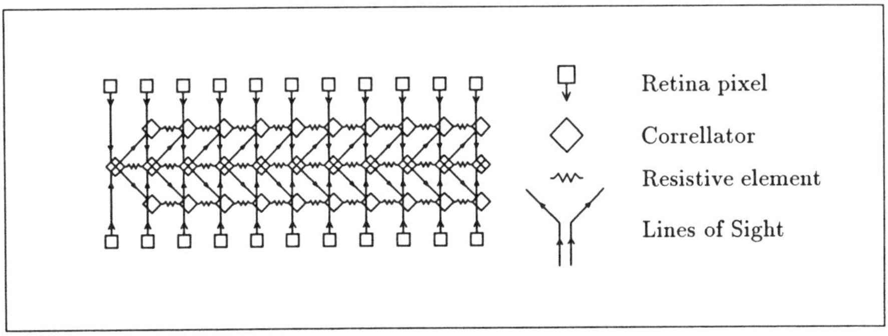

Figure 4 Chip architecture. This schematic is topologically identical to the representational framework of the algorithm shown in Figure 3 except fewer pixels and disparity planes are shown. The top and bottom rows of pixels are one- dimensional retinas. Between the retinas is the array of disparity planes. The central row represents zero disparity. The computational elements shown in the legend are described in the text. The actual chips have 40 pixel retinas and nine disparity planes.

The heart of the chip is the correlator array. The input to a correlator comes from two pixels, one pixel from each retina. The retinal output is an analog function of the image. The current input to a correlator is a nonlinear combination of the output from two pixels. Each disparity plane represents a point-by-point cross-correlation between the two retinas, at a spatial offset corresponding to that plane's disparity.

We have fabricated and characterized two variants of the chip. The original version of the chip uses static contrast as a feature to drive the correlation. The computation performed by the retinas is similar to a convolution of the image with a center-surround filter. The analog values of the computation are used to compute a cross-correlation between the two images. The retinas are one-dimensional versions of the silicon retina (Mahowald, 1989). The photoreceptor takes the logarithm of the incoming light intensity. Outputs of the photoreceptors are averaged on a resistive network. The differential voltage between the resistive net and the photoreceptor provides input to a modified Gilbert multiplier, which is located in a correlator cell (Mead, 1989). The multiplier produces a current that is the basis for the correlator computation.

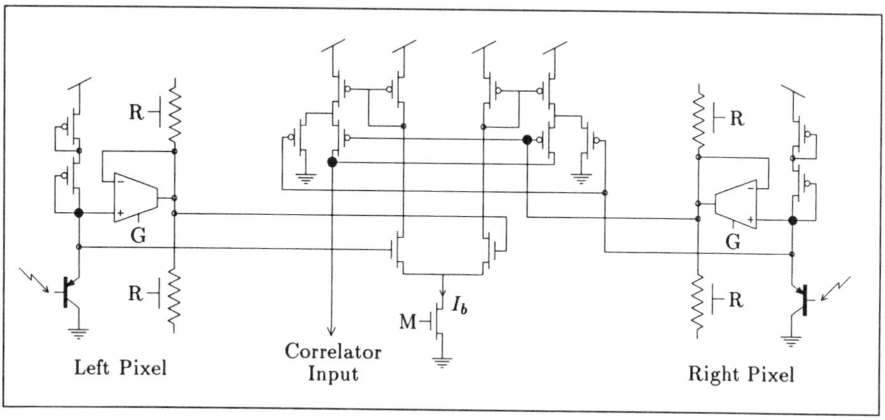

Figure 5 Retinal circuitry and modified Gilbert multiplier. The left pixel is located in the left retina, and the right pixel is located in the right retina. The multiplier is located in a correlator cell. Each correlator has its own multiplier, which receives input from a unique combination of pixels.

The multiplier performs a four quadrant multiplication biased around $\frac{I_b}{2}$:

$$I_{\text{mult}} = (\frac{I_b}{2})(1 + \tanh(V_r)\tanh(V_l)).$$

V_r and V_l are the differential voltages from the right and left retinas. The resistive network acts as a reference level for the computation; intensities brighter than the average are "white," whereas those darker than the average are "black." The output of a black pixel multiplied by the output of another black pixel results in a current into the correlator that is greater than $\frac{I_b}{2}$, whereas the output of a black pixel multiplied by the output of a white pixel results in a current that is less than $\frac{I_b}{2}$. The magnitude of the signal coming into the multiplier is related to the contrast of the feature.

Another variant of the chip uses time derivatives of intensity as features for the correlation. The retinas consist of arrays of change-sensitive pixels, called *hysteretic receptors* (Delbrück and Mead, 1989). The response of one of these pixels to an increase in light intensity is a sharp and transient decrease in the output voltage. The response to a decrease in light intensity is very small. The hysteretic receptor has a gain for transients that is about 100 times larger than the gain for the steady-state illumination level.

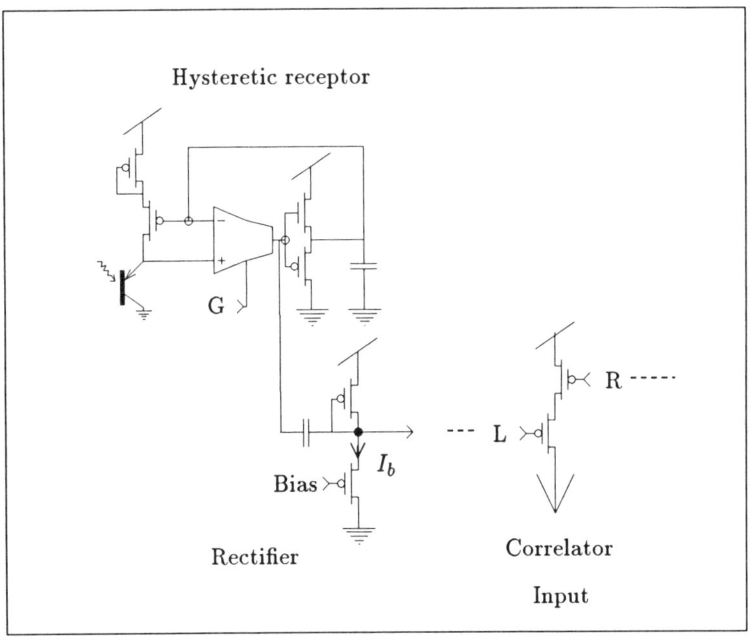

Figure 6 Time derivative pixel and series connected transistors. The half-wave current-rectifier circuit bias current is labeled I_b.

The output of the hysteretic element is capacitively coupled to a half-wave current-rectifier circuit biased to generate a small current when the intensity is unchanging (Lazzaro, this volume). This bias also controls the time constant of the circuit. Even small increases in intensity increase the current through the rectifier. The voltage output of the rectifier is broadcast to the correlator array along a line of sight.

The correlation of signals from the right and left pixels is performed in the correlator cell by means of two serially connected transistors. In subthreshold, these two transistors compute the function:

$$I_{\text{mult}} = \frac{I_\text{r} I_\text{l}}{I_\text{r} + I_\text{l}}.$$

I_r and I_l are the currents through the rectifiers in the right and left pixels, respectively. This operation is a normalized multiplication of the two retinal inputs. If either retinal input is small, the current into the correlator is small. When the images on the retinas are unchanging, the input to the correlator is proportional to the bias current of the rectifiers. When there is a change in the input, the current into the correlator can be up to three orders of magnitude larger than the bias current (Figure 7).

The time-derivative chip has the advantage that the correlator cell is compact. In addition, the derivative-based correlator input has a range larger than the multiplier current in the static contrast chip. However, the operation of the correlator array is easier to conceptualize in the chip that uses static contrast as a matching primitive. Most of our discussion will focus on the operation of the static contrast-sensitive chip; description of the change-sensitive chip will be postponed until the section on experimental results.

The operation of the correlator array depends on cooperative interactions among correlators. The connections between correlators implement the continuity and uniqueness constraints of the original algorithm. Figure 8 schematically depicts the interactions at a correlation element. The correlator itself is simple; it is a single electrical node. The computations performed at a correlator node are complex: Currents from saturating nonlinear sources are summed to create a voltage; this voltage is used to control a nonlinear conductance; and the conductance determines the extent of electrotonic coupling. Although the number of interactions at this node are small by neural standards, the correlator circuit demonstrates the high computational density available in an analog medium.

The uniqueness constraint is implemented via a winner-take-all (WTA) circuit shown in Figure 9 (Lazzaro *et al.* 1989). The WTA circuit establishes a competitive feedback interaction between all the correlators along a line of sight. Each correlator will participate in two competitive groups; one group for a pixel from the left retina and another group for a pixel from the right retina. On the stereo correspondence chip, the number of channels participating in each WTA competition is equal to the number of disparity planes.

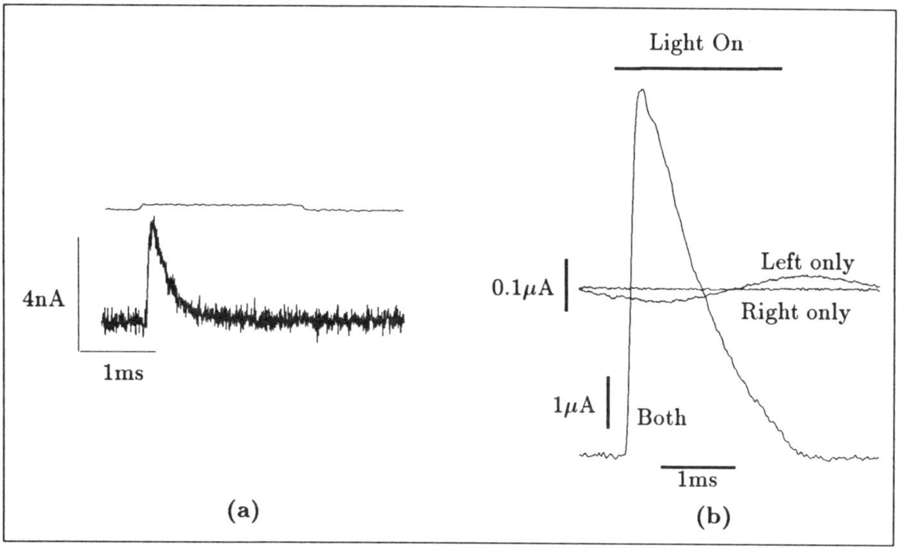

Figure 7 The retinal response and correlation performed by the motion-sensitive Marr chip. **(a)** The retinal signal generated by a pixel in response to a flashing LED. The LED signal is shown in the top trace. The magnitude of the modulation is 5 percent. The output of the pixel is shown in the bottom trace. The current maximum is approximately 4 nA. The zero input bias into the correlator is 100 pA. **(b)** Response of a single correlator to a flashing LED. The traces labeled "Right Only " and "Left Only" show the output of the correlator when only one retina is stimulated. The trace labeled "Both" shows the correlator response when both retinas are stimulated.

The competitive interaction is optimized to suppress false matches. Because competition takes place along lines of sight, all false matches are competing with true matches. A simple analysis of the current flowing through the WTA circuits of the correlators receiving retinal inputs shows that false matches can never suppress true matches.

The matching process can be broken down into competitions between two targets at a time. Imagine that there are two targets, A and B. Denote the pixel response to target A as a and the pixel response to target B as b. Figure 10(a) shows the system response to a two target stimulus in which $a > b$. Input to the correlator that correlates the response of pixel a_r in the right retina and pixel a_l

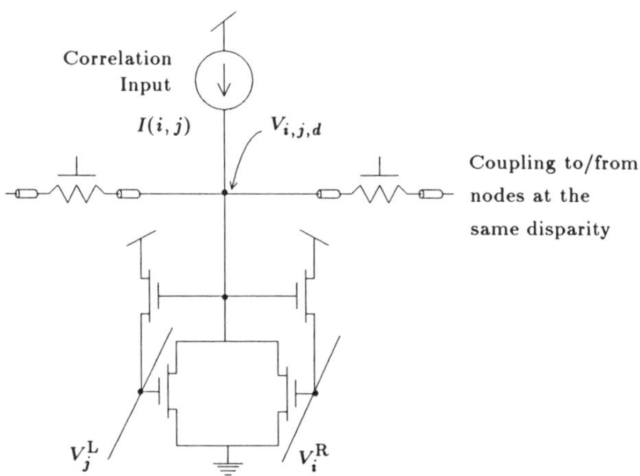

Figure 8 Simplified version of the electrical interactions at a correlation element. The correlator output voltage, $V_{i,j,d}$, is determined by the sum of the currents flowing into the node. The primary input, $I(i,j)$, is a current that is a nonlinear AND–type function of the signals from pixel i in the right retina and pixel j in the left retina. Correlation elements in the same disparity plane are coupled by resistive elements. The schematic of the resistive element indicates that the resistance is nonlinear and controllable. Two transistors drawing current to ground provide inhibition. The voltages V_i^R and V_j^L are a function of the correlator voltages along the line of sight of pixel i and pixel j respectively.

in the left retina is a^2. If $a > b$, this correlator sets the WTA circuits associated with a_r and a_l, so that each WTA circuit sinks a current $\frac{a^2}{2}$ (a total of a^2). The retinal input to the false match between a_r and b_l is ab. The current available for this false match to suppress the true match between b_r and b_l is $ab - \frac{a^2}{2}$. The false match can suppress the true match only if $ab - \frac{a^2}{2} > \frac{b^2}{2}$. Figure 10(b) shows the amount of current by which the false match can suppress the true match, as a function of the ratio of the contrast of the targets, $\frac{a}{b}$. Intuitively, the false matches AB have a higher correlation input than the true match BB, but the suppression of the AB matches by the AA match is always sufficient to let the BB match win. The false match never has enough current to win

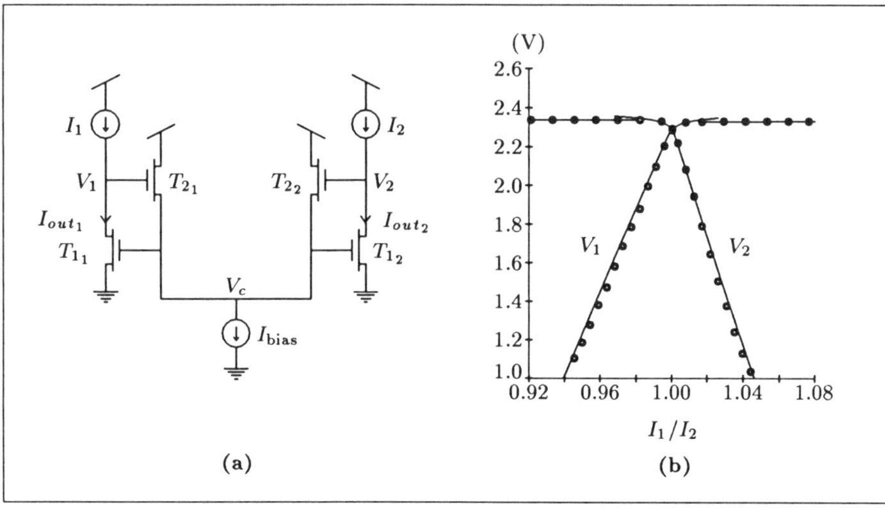

Figure 9 The winner-take-all circuit. **(a)** Schematic of a simple two-channel winner-take-all circuit. To understand how the circuit works in subthreshold, imagine that the circuit is in equilibrium and that each channel is receiving an identical input current. In this configuration, $I_1 = I_2 = I_{out_1} = I_{out_2}$. The voltage on the common line, V_c, is therefore constrained to be logarithmic in the input current. The voltages on the output nodes, V_1 and V_2, are constrained to supply the bias current to the common line through source-follower transistors, T_{2_1} and T_{2_2}. These voltages are above the common line voltage by an amount that is logarithmic in the bias current. To make one channel win over the other, we increase its input current. Increasing the current to one channel charges up that channel's output node. The voltage on the common line will follow the output voltage of the winning channel with a voltage difference set by the bias current. The output node will stop charging when the current through its T_1 transistor is equal to the new input current. The output voltage of the winning channel will increase logarithmically with input current. The loser will be suppressed. **(b)** Current-voltage characteristic of the two channel WTA circuit. The voltage output of the two channels is plotted against the ratio of their input currents. Since the voltage on the common line, V_c, controls the current out of both channels, the capacitor of the channel with less current will be discharged until its T_1 transistor draws only its input current. For current differences between the channels of more than a few percent, the T_1 transistor of the losing channel will come out of saturation; the output voltage will be within a few $\frac{kT}{q}$ of ground. When the current difference between channels is small, the output voltage on the losing channel is determined by the Early voltage of the T_1 transistor and by the level of the input current.

over the true match. Correlations between two features of equal value present the most difficult case. False matches then generate the same correlator inputs

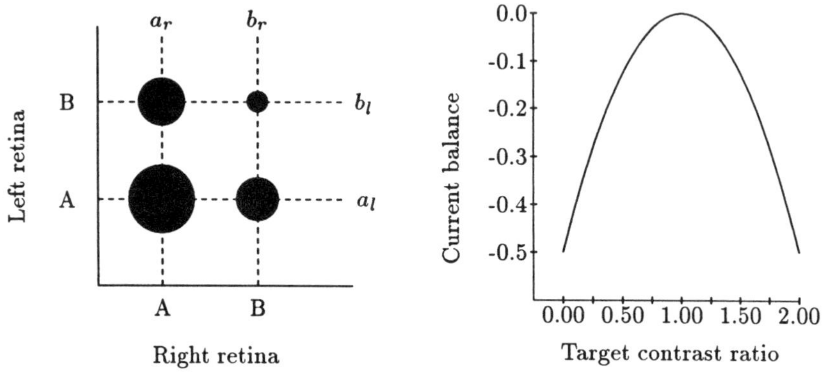

Figure 10 Suppressing false targets. **(a)** The computation of the correspondence between two targets, A and B, is illustrated using the representation explained in Figure 3. The magnitude of the retinal input to the correlator is shown as the area of a filled circle. Pixels associated with the lines of sight are labeled. **(b)** The function $f(\alpha) = -\frac{1}{2}\alpha^2 + \alpha - \frac{1}{2}$, where α is the ratio of the contrast of the targets, $\frac{a}{b}$. The function $f(\alpha)$ represents the amount of current by which the false match can suppress the true match in units of b^2. If the value of $f(\alpha)$ is negative, that much extra current would need to be supplied to the false match in order for it to tie with the true match in the WTA competition.

as do true matches. We can suppress these false matches only by invoking the continuity constraint.

The continuity constraint is mediated by resistive elements that couple the correlation nodes within a disparity plane. The resistor is implemented with the same circuit that is used in the static retina to compute the local average illumination level. The resistor circuit has the saturating current-voltage characteristic described by

$$I = G \tanh(\frac{V}{2}).$$

The units of voltage are $\frac{kT}{q}$. The parameter G is controlled by an external voltage.

In the correlator array, resistive coupling locally averages the activity level within a disparity plane by providing a path for current flow. Within a uniform disparity region of the image, the correlators at the correct disparity plane will all be receiving positively correlated retinal input. Other disparity planes will be receiving some anticorrelated input. A correlator that is receiving anticorrelated input will draw current from neighboring correlators within its disparity plane. The WTA circuit is thus able to suppresses the correlators in the disparity planes signaling false targets.

The strength of coupling must be limited if the chip is to function properly. The saturating characteristic of the resistor is important in allowing the disparity to change as a function of horizontal image position. The current that can be drawn from the winning disparity region across a disparity discontinuity is limited by the nonlinearity of the resistor. Saturation of the resistor allows a large voltage difference to form across the edge (Hutchinson, 1988).

We can write a system of equations that capture most of the operation of the system. These equations assume that the transistors in the circuits are being operated in subthreshold. The drain–source current $I_{\rm ds}$ of a transistor operated in subthreshold is given by:

$$I_{\rm ds} = I_0 e^{\kappa V_{\rm g} - V_{\rm s}} (1 - e^{V_{\rm d} - V_{\rm s}} + \frac{V_{\rm d} - V_{\rm s}}{V_0})$$

where $V_{\rm g}$ is the gate voltage, $V_{\rm s}$ is the source voltage, and $V_{\rm d}$ is the drain voltage. The constant I_0 is about 10^{-15} amps. The constant κ represents the body effect (Mead, 1989) and is about 0.7. The constant V_0 is the Early voltage and is typically a few tens of volts (Mead, 1989). All voltages are in units of $\frac{kT}{q}$.

The dynamical equation describing the interactions at a correlator node is

$$\begin{aligned} C\frac{d}{dt}V_{i,j,d} &= I(I_i^R, I_j^L) \\ &+ G\tanh\Big(V_{i+1,j+1,d} - V_{i,j,d}\Big) \\ &+ G\tanh\Big(V_{i-1,j-1,d} - V_{i,j,d}\Big) \\ &- I_0 e^{\kappa V_i^R}\left(1 - e^{-V_{i,j,d}} + \frac{V_{i,j,d}}{V_0}\right) \\ &- I_0 e^{\kappa V_j^L}\left(1 - e^{-V_{i,j,d}} + \frac{V_{i,j,d}}{V_0}\right) \end{aligned} .$$

In this equation, C is the capacitance on a correlator node. $V_{i,j,d}$ is the correlator voltage, and the subscripts refer to the inputs from pixel i on the right retina, pixel j from the left retina, and disparity plane d. The disparity planes run from $-d_0$ to $+d_0$, zero disparity referring to alignment of the two retinas. $I(I_i^R, I_j^L)$ is the correlation input from the right and left pixels. The next two terms are the resistor currents to and from neighboring correlators at the same disparity. The constant G is controllable externally. The final two terms represent the nonlinear inhibition performed by the WTA circuit.

The common-line voltages on the winner-take-all circuits are given by

$$V_i^R = \ln\left(\sum_{d=-d_0}^{d=+d_0} \frac{I_0}{I_b} e^{\kappa V_{i,i-d,d}}\right)$$

for common lines emmanating from the right retina. A similar expression holds for the left retina. This equation represents a summation over a line of sight. The WTA bias current is denoted I_b.

It is difficult to gain a quantitative understanding of this system, since the circuit elements are nonlinear and extensively cross-coupled. For a given stimulus, the total positive current into each disparity plane is a fixed quantity, depending only on the retinal stimulus and not on the state of the correlator array. Current can only leave the disparity plane through the WTA circuits. The current–voltage characteristic of the WTA circuit determines the correlator voltage. The distribution of correlation input among the correlators in a disparity plane determines the state of the system. The spread of current within a single disparity plane is a function of the voltages of the correlators in the other disparity planes. In a simple resistive network, the space constant is set by the relationship between the lateral resistance and the conductance to ground. In the correlator array, the WTA inhibition sets the strengths of conductances between the output nodes and ground, based on the maximum correlator voltages along each line of sight. Consequently, the space constant of the network depends on the data. A current may propagate laterally for a long distance in one context, yet the same current may be quickly shunted to ground under another set of inputs. Context dependence makes the circuit (and the algorithm (Marr *et al.* 1978)) very difficult to characterize. This report describes the function of the chip by looking at specific examples.

EXPERIMENTAL RESULTS

Presentation of data to the system is simplified by the fact that photosensors are integrated on the chip; a single lens focused on the surface of the silicon projects an image onto two parallel, one-dimensional retinas. We can generate artificial disparities by using two images of bars, one for each retina. When the two images are identical, the shift of one image relative to the other determines the disparity. We can create images with multiple disparity regions by shifting portions of the images relative to each other. The output of the retinas and of the two dimensional array of correlators is scanned serially off the chip using the method described in (Sivilotti *et al.* 1987). The output voltages of the correlators produce an output current by controlling the gate of a transistor. The output current is sensed by an off-chip current-sense amplifier.

The response of the chip to a simple pattern with two disparity regions is shown in Figure 11. The image is clearly segmented into two depth planes. In this example, the correlation is performed based on dense retinal signals. The space constant of the resistive network in the retina is set to be large, so that its principle function is to act as the reference value for determining positive and negative contrast. Because the stimulus is high contrast, most of the retinal outputs are saturated either high or low. Many false matches occur, since several consecutive pixels on the retinas are black or white.

To limit the number of false targets generated by large areas of uniform contrast, we can use the retina to enhance edges in the image. Figure 12 shows the chip's response to an edge-enhanced input. The computed disparity of the wide bar is continuous, even though the input is present mainly at the edges. When all the disparity planes are tied at a low value due to lack of retinal input, the current through the resistor can spread along a disparity plane to fill in the correct solution. When the disparity planes have similar retinal inputs, the impedance of a losing correlator node is set by the Early voltage of its T_1 transistor. If the lateral resistance is small, the space constant of the losing disparity plane is large, so the solution spreads a long way.

In addition to filling in the solution in regions of low retinal input, the resistors help to suppress false matches. Ideally a false match can never be bigger than a true match, so the resistors do not have to draw much current for the false matches to be suppressed. Unfortunately, the circuit elements we are using are not ideal. Transistor mismatches in the multipliers and pixel elements introduce random variations in the correlator inputs. Offsets may result in a false match being up to twice as big as a true match (Mead, 1989). The resistive coupling between correlation nodes helps reduce the effect of circuit offsets.

The ability of the resistors to suppress unwanted signals is a function of their strength. Decreasing the resistance couples correlators within a disparity plane more strongly and helps to suppress false matches and offsets. The strength of the resistors determines the area over which retinal inputs are averaged. The disparity plane with the largest average input wins. If the correlators within a disparity plane are too strongly coupled, then they act as a unit. The retinal input to the entire disparity plane is averaged, and the plane with the largest total input current wins. To allow breaks in disparity, we must limit the resistance.

We can investigate the effects of changing the coupling between correlators by examining the area around the discontinuity, in an image with two disparities. Figure 13 shows the output of the correlators on both of the winning disparity planes. When the lateral resistance is large, very little current flows,

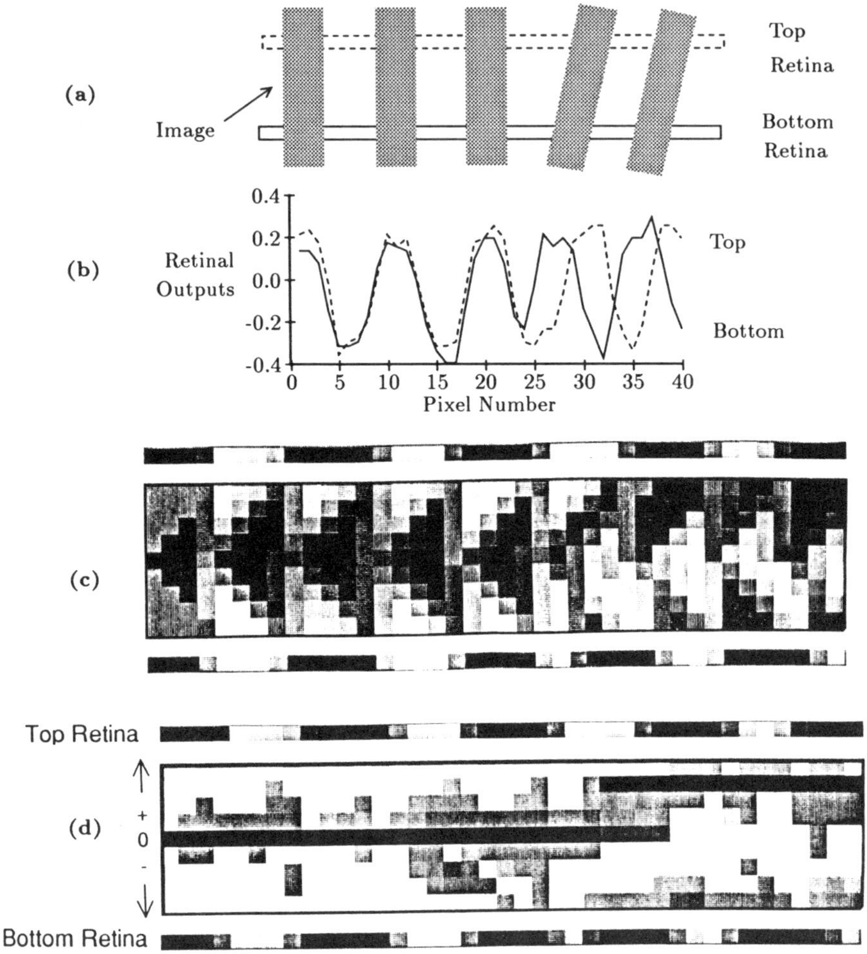

Figure 11 Finding two disparity planes. **(a)** The chip was tested using black and white bar patterns that could be tilted, as shown by the shaded bars. The degree of tilt determines the disparity of the pattern between the two retinas; when the bars are perpendicular to the retinas they are at zero disparity. We can construct a disparity discontinuity by combining two such bar patterns at different degrees of tilt. Such a stimulus is shown in dotted outline superimposed over the array. **(b)** Retinal response to this input pattern. The space constant of the resistive network is longer than the bar width, so the retinal output is not edge enhanced. **(c)** The product of the two retinal outputs, computed off-chip and displayed in the same format as the actual chip output. It is an approximation to what the array of correlators is receiving as input. High correlation values are dark. High correlations away from the correct disparity plane are false matches. **(d)** Analog voltage output from the chip, encoded by gray levels, shows segmentation into two disparity planes. The cooperative interactions among correlators suppresses false correlations.

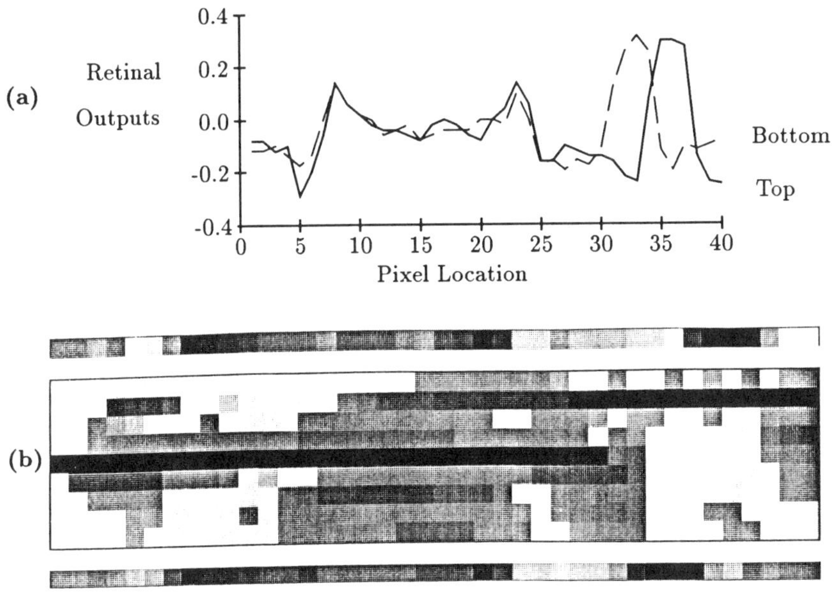

Figure 12 Filling in the solution. **(a)** Retinal response to this input pattern. The space constant of the resistive network is shorter than the width of the large bar. Due to the averaging properties of the resistive net, the response to the edge of the wide bar is smaller than the response to the narrow bar. The response is enhanced at the edge of the wide bar. In the central region of the wide bar, the retinal output is near zero. **(b)** Analog output from the chip. The solution breaks into two disparity regions. The central area of the bar is filled in.

and the winning correlator clearly dominates the loser. Since the resistor saturates, the voltage difference across the break can increase without drawing more current across the edge.

As the resistance decreases relative to the multiplier bias, the disparity discontinuity becomes less sharp. The resistors draw current from the winning correlator at the break. Figure 13 illustrates the effect of decreases in the lateral resistance. When the ratio of the multiplier bias current to the resistor saturation current is 0.30 (small resistance case), the current flowing through the resistor is able to supply some current to all the correlators across the break. The solution propogates across the change in disparity, so the winning disparity

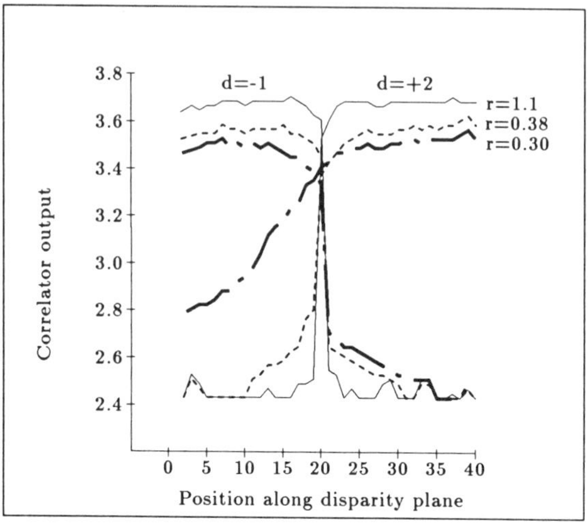

Figure 13 Correlator outputs for an image with two regions of different disparity. The input image has one region at disparity +2 next to another region of disparity -1. The boundary between the regions occurs in the center of the image. The analog outputs of the correlators on the +2 and -1 disparity planes are shown for different multiplier bias currents. The off-chip current-sense amplifier saturates at 3.7 volts. The zero-point reference voltage is 2.4 volts. The ratio, r, of the multiplier bias current to the resistor bias current is shown next to the curve, along with the disparity plane, d.

plane is less distinguishable from the losing plane. When the ratio is larger, the losing plane is clearly distinguished from the winning plane.

The variant of the chip that uses motion signals as features for correlation has a larger input signal and smaller offsets than has the static contrast-sensitive chip. There is only one current mirror between the output voltage generated by the retina and the input current to the correlator. These mirrors are mismatched between correlators, but the mismatches are smaller than those generated by the cascaded mirrors in the Gilbert multiplier. Because of its stronger input signals, the time derivative chip is more able to fill in areas of the solution that are not receiving retinal input, while still finding multiple disparity regions. Using a 40-pixel input array, the time derivative chip is able reliably to find correct solutions for images with three disparity regions, whereas the static contrast chip can discriminate only two disparity regions (Figures 14, 15 and 16). The larger correlation input, however, means that the correlators are less able to suppress transient false matches (Figure 15).

For images with uniform intensity areas, the motion chip eliminates many false targets while maintaining a large input signal. The spatial extent of the time-derivative signal coming from the retinas is set by the speed of motion and the time constant of the derivative elements. An element generates few false matches by correlating with its previous position (rather than its current position) in the other retina if the time constant is short. As in the case of the edge-enhancing static retina, the retinal signal is small in areas of uniform illumination. Spatially uniform intensity regions do not generate time-derivative signals when the image is displaced. The retinal input to the correlators generated at moving edges is very large compared to the retinal input in regions of unchanging intensity. The current supplied by a saturated resistor can be large compared to the static correlator input, but still small compared to the motion-generated signals. The solution can propagate effectively in regions of no image motion, whereas the signals from moving edges are easily able to break the solution into multiple disparity regions.

The integrative properties of the correlators may aid in filling in the solution. The dynamics of the correlators are set by the WTA circuits. The winning channel is discharged logarithmically in time by what is essentially a diode-connected transistor. In the absence of any competing input, the winning channel takes a long time to decay (Figure 16). Therefore, the solution is integrated spatially as the stimulus moves over the pixel array. The fact that the winner stays active for a long time does not generate any false matches. In contrast, many false matches would be generated if the retinal derivative circuit had a large time constant.

Figure 14 Image with three constant-disparity regions. Boundaries between regions are indicated by vertical markers. This binocular pair, when fused, shows a central region containing 4 white bars standing out over a surrounding background.

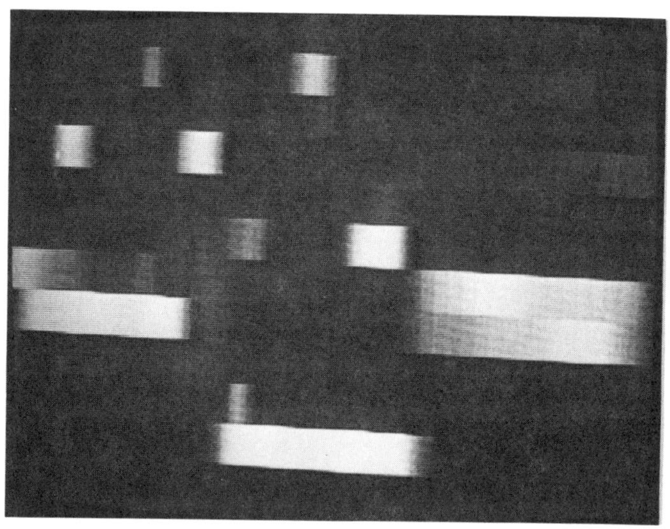

Figure 15 Scanned correlator-array output for motion sensitive chip, showing solution divided into three regions. Image is being moved slowly across the retinas. Bright areas signify a match. The picture was taken 0.25 seconds after the start of image motion. Transient false matches are not well suppressed.

(a)

(b)

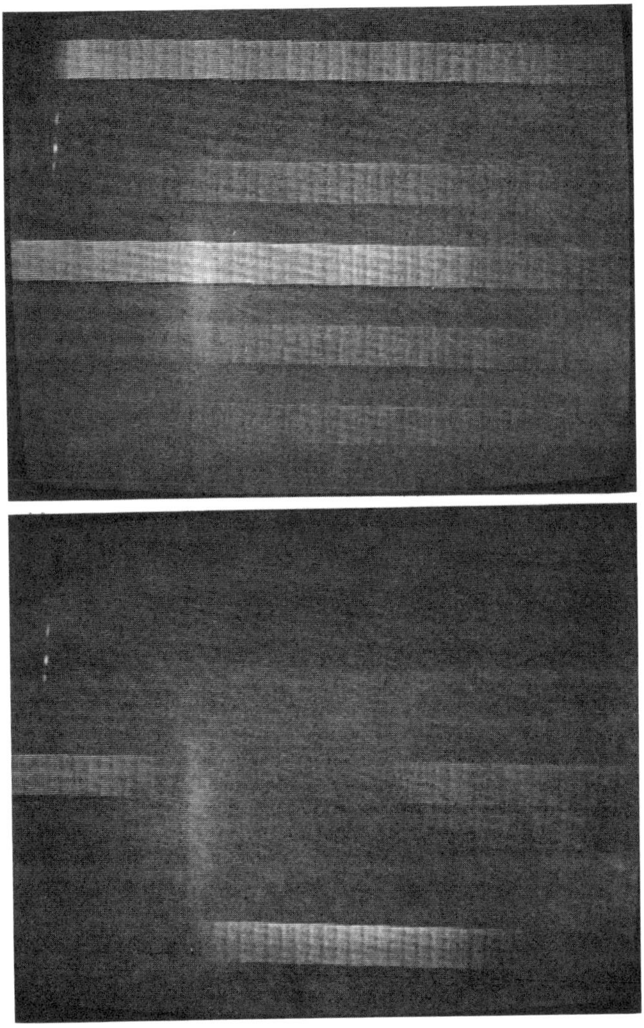

Figure 16 (**a**) Scanned correlator array output showing response of the chip to the still input pattern. When there are no motion signals from the retina, the correlator array outputs are averaged over the whole disparity plane. Each disparity plane appears as a horizontal bar. (**b**) Scanned correlator array output, 0.65 seconds after stopping motion of the input pattern. The solution is still visible because the state of the winning correlators decays slowly.

DISCUSSION

Computational and synthetic approaches to artificial vision have a synergistic relationship. The stereo matching circuit profits from many of the strengths of Marr's computational algorithm. For example, resistive coupling between correlators enables the system to function in spite of large offsets. To average out offsets and to suppress false matches, we must set the resistors at a low value. Strong coupling results in a limited number of disparity regions that can be correctly identified in a single image. It is not yet understood what the intrinsic limitations are on the number of different disparity planes the chip can find. That the chip can find any solution in the presence of large offsets is a testament to the robust nature of the collective algorithm.

The chip architecture, which is an embodiment of the representational framework of the algorithm, uses a value-unit encoding (Ballard, 1986). Rather than encoding target depth by the magnitude of a voltage, activation of a correlator unit represents the presence of a target in a discrete region of three-dimensional space. Disparity is determined by the pattern of activity in the correlator array. The WTA inhibition is analog in nature; if two correlation elements are more or less equally stimulated by retinal input and cooperative interactions, then their outputs are comparable. If the disparity of the image is between two disparity planes, both are activated. When properly interpreted, the value-unit encoding achieves a resolution that is finer than a single pixel, even though the individual circuit elements are imprecise.

Implementation of algorithms in a physical medium stimulates their development in unforseen directions. For example, the use of optically acquired images precipitated the development of a variety of input features for stereo matching (Nishihara, 1984). These features are more robust to noise and vertical offsets between images than are the abstract binary tokens used as matching features in the original algorithm.

The static contrast-sensitive chip uses the analog value of a center-surround computation, which is easy to implement, as a primitive for stereo matching. The computation performed by the retinas on the chip retains information about the contrast between objects and is biologically plausible (Mahowald and Mead, 1989). The use of a center-surround matching primitive has been investigated by Mayhew and Frisby (1981). The use of time derivatives of intensity as a matching primitive has not been previously explored. Retinal neurons that generate transient responses, similar to those generated by the time-derivative silicon retinas, are known to project, via several relays, to disparity-tuned cells in layer IVb of the primary visual cortex (Poggio, 1984). The use of transient matching primitives introduces time as a representational dimension in the stereo correspondence computation.

Little research has been done in time-based algorithms for stereopsis because it is difficult to simulate temporal functions using traditional methods. It is known that time is an intrinsic part of the disparity computation in natural systems. Perceptual psychologists have shown that binocular time delay and disparity can be substituted for each other in moving stimuli (Burr and Ross, 1979). Binocular time delay has been used to characterize disparity sensitive neurons in visual cortex (Gardner *et al.* 1985). Signals that are time delayed between the two eyes result from motion in a complex environment in which surfaces occlude one another. (Shimojo *et al.* 1985).

Parallel, analog hardware greatly facilitates real-time processing of complex inputs. The time-derivative correlating chip is a first step toward a real-time, interactive system for stereopsis. Using time derivatives has improved the performance of the chip by reducing the effect of offsets. In the future, we hope to expand the chip's representation to include binocular time delay. By taking advantage of the properties of the analog medium, we hope to gain further insights into the problem of stereopsis.

Acknowledgements

We thank Jin Luo and Tor Lande for technical assistance. We thank Hewlett-Packard for computing support, and DARPA and MOSIS for chip fabrication. This work was sponsored by the Office of Naval Research and the System Development Foundation.

References

Ballard, D.H. (1986). Cortical connections and parallel processing: Structure and function. *The Behavioral and Brain Sciences* **9:** 67–120.

Burr, D.C., and Ross, J. (1979). How does binocular delay give information about depth? *Vision Research* **19:** 523–532.

Delbrück, T., and Mead, C.A. (1989). An electronic photoreceptor sensitive to small changes in intensity. In Touretsky, D. S. (ed), *Advances in Neural Information Processing Systems* 1., pp 712–727, San Mateo, CA: Morgan Kaufman.

Gardner, J.C., Douglas, R.M., and Cyander, M.S. (1985). A time-based stereoscopic depth mechanism in the visual cortex. *Brain Research* **328:** 154–157.

Hutchinson, J., Koch, C., Luo, J., and Mead, C. (1988). Computing motion using analog and binary resistive networks. *IEEE Computer* March pp. 52–63.

Julesz, B. (1960). Binocular depth perception of computer-generated patterns. *Bell Syst. Tech. J.* **39:** 1125–1162.

Julesz, B. (1971). *Foundations of Cyclopean Perception.* Chicago, IL: The University of Chicago Press.

Lazzaro, J., Ryckebusch S., Mahowald, M.A., and Mead, C.A. (1989). Winner-Take-All circuits of $O(n)$ complexity. In Touretsky, D.S. (ed), *Advances in Neural Information Processing Systems* 1. pp. 703–711, San Mateo, CA: Morgan Kaufman.

Mahowald, M.A. and Mead, C.A. (1989). Silicon retina. In Mead, C.A. *Analog VLSI and Neural Systems*, pp. 257–278, Reading, MA: Addison-Wesley.

Marr, D., Palm, G., and Poggio, T. (1978). Analysis of a cooperative stereo algorithm. *Biological Cybernetics* **28:** 223–239.

Marr, D., and Poggio, T. (1976). Cooperative computation of stereo disparity. *Science* **194:** 283–287.

Marr, D. (1982). *Vision*, New York: W. H. Freeman.

Mayhew, J. and Frisby, J.P. (1981). Psychophysical and computational studies towards a theory of human stereopsis. *Artificial Intelligence* **17:** 349–385.

Mead, C.A. (1989). *Analog VLSI and Neural Systems.* Reading, MA: Addison-Wesley.

Mead, C.A. and Mahowald, M.A. (1988). A silicon model of early visual processing. *Neural Networks.* **1:** 91–97.

Nishihara, H. (1984). Practical real-time imaging stereo matcher. *Optical Engineering* **23:** 536–545.

Poggio, G. (1984). Processing of stereoscopic information in primate visual cortex. In Edelman, G. M., Gall W. E., and Cowan, W. M. (eds), *Dynamic aspects of neocortical function*, pp. 613–635, New York: John Wiley & Sons.

Poggio, G. and Poggio, T. (1984). The analysis of stereopsis. *Annual Review of Neuroscience* **7:** 379–412.

Shimojo, S., Silverman, G.H., and Nakayama, K. (1985). An occlusion-related mechanism of depth perception based on motion and interocular sequence. *Nature* **333:** 265–268.

Sivilotti, M.A., Mahowald, M.A., and Mead, C.A. (1987). Real-time visual computations using analog CMOS processing arrays. In Losleben, (ed), *Advanced Research in VLSI, Proceedings of the 1987 Stanford Conference,* pp. 295–312, Cambridge, MA : MIT Press.

10

ADAPTIVE RETINA

Carver Mead
California Institute of Technology
Pasadena, California, 91125

Retinal Computation

Mahowald describes a silicon model of the computation performed by the first layer of visual processing, located in the outer plexiform layer of the retina [Mahowald 88,89]. The lateral spread of information at the outer plexiform layer is mediated by a two-dimensional resistive network. The voltage at every point in the network represents a spatially weighted average of the photoreceptor inputs. The farther away an input is from a point in the network, the less weight it is given. The weighting function decreases in a generally exponential manner with distance.

Each photoreceptor in the network is linked to its six neighbors with resistive elements, to form the hexagonal array shown in Figure 1. Each node of the array has a single bias circuit to control the strength of the six associated resistive connections. The photoreceptors act as voltage inputs that drive the resistive network through conductances. Because a transconductance amplifier was used in place of a bidirectional conductance, the photoreceptor acts an effective voltage source. No current can be drawn from the output node of the photoreceptor, because the amplifier input is connected to only the gate of a transistor.

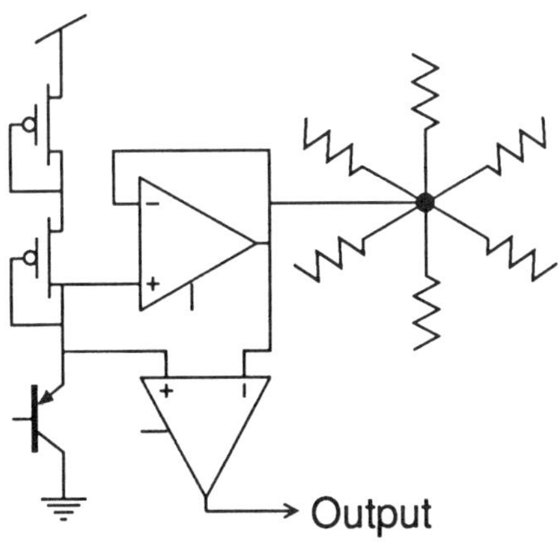

Figure 1 Schematic of pixel from the Mahowald retina. The output is the difference between the potential of the local receptor and that of the resistive network. The network computes a weighted average over neighboring pixels

The resistive network computes a spatially weighted average of photoreceptor inputs. The spatial scale of the weighting function is determined by the product of the lateral resistance and the conductance coupling the photoreceptors into the network. Varying the conductance of the transconductance amplifier or the strength of the resistors changes the space constant of the network, and thus changes the effective area over which signals are averaged.

The receptive field of the output of this computation shows an antagonistic center-surround response. This behavior is a result of the interaction of the photoreceptors, the resistive network, and the output amplifier. A transconductance amplifier provides a conductance through which the resistive network is driven toward the photoreceptor potential. A second amplifier senses the voltage difference across the conductance, and generates an output proportional to the difference between the photoreceptor potential and the network potential at that location. The output thus represents the difference between a center intensity and a weighted average of the intensities of surrounding points in the image.

Frank Werblin suggested that the model shown in Figure 1 might benefit from the known feedback connections from resistive network to photoreceptor circuit. Our first attempt to incorporate this suggestion is shown in Figure 2. In this pixel circuit, the output node is the emitter of the phototransistor. The current out of this node is thus set by the local incident light intensity. The current into the output node is set by the potential on the resistive network, and hence by the weighted average of the light intensity in the neighborhood. The difference between these two currents is converted into a voltage by the effective resistance of the output node, determined primarily by the Early effect. The advantage of this circuit is that small differences between center intensity and surround intensity are translated into large output voltages, but the large dynamic range of operation is preserved. A retina fabricated with this pixel did indeed show high gain, and operated properly over many orders of magnitude in illumination. The transconductance amplifier has a hyperbolic-tangent relationship between output current and input differential voltage. For proper operation, the conductance formed by this amplifier must be considerably smaller than that of the resistive network node. For that reason, when a local output node voltage is very different from the local network voltage, the amplifier saturates and supplies a fixed current to the node. The arrangement thus creates a center-surround response of only slightly different form from that of the straightforward implementation of Figure 1.

Once the new circuit was operating, it was immediately clear that its higher gain made it much more sensitive to transistor offset voltages than lower-gain versions had been. Under uniform illumination, most pixel outputs were driven to one rail or the other. Of course biological retinas must have precisely the same problem. No two receptors have the same sensitivity, and no two synapses have the same strength. The problem in wetware is even more acute than it is

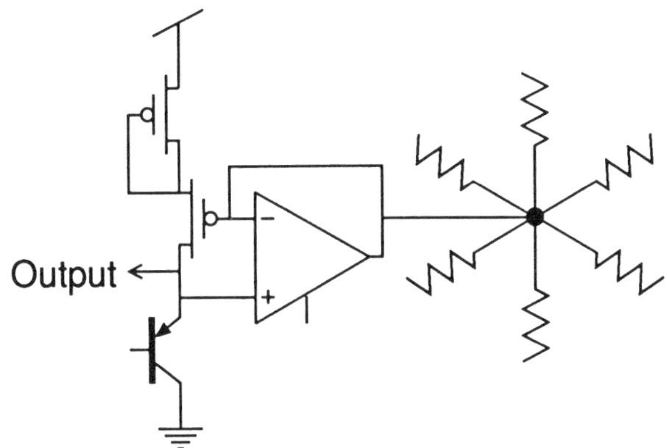

Figure 2 Schematic of a simplified pixel circuit that performs an operation similar to that performed by the Mahowald unit of Figure 1.

in silicon. It is also clear that biological systems use adaptive mechanisms to compensate for their lack of precision. The resulting system performance is well beyond that of our most advanced engineering marvels. Well, if biology can do it, so can we. Once we understand the principle, we can incorporate it into our silicon retina. Before we can even start, we need an adaptive mechanism.

Adaptive Mechanism

All our analog chips are fabricated in silicon-gate CMOS technology [Mead 89]. If no metal contact is made to the gate of a particular transistor, that gate will be completely surrounded by silicon dioxide—the world's best insulator. Any charge parked on such a floating gate will remain for eons. The first floating-gate experiments of which I am aware were performed at Fairchild Research Laboratories in the mid 1960's. The first product to represent data by charges stored on a floating gate was reported in 1971 [Frohman 71]. In this device, which today is called an EPROM, electrons are placed on the gate by an avalanche breakdown of the drain junction of the transistor. This injection can be done selectively, one junction at a time. Electrons can be removed by ultraviolet light incident on the chip. This so-called erase operation is performed on all devices simultaneously. In 1985, Glasser reported a circuit in which either a binary one or a binary zero could be stored selectively in each location of a floating-gate digital memory [Glasser 85]. The essential insight contributed by Glasser's work was that there is no fundamemtal assymetry to the current flowing through a thin layer of oxide. Electrons are excited into the conduction band of the oxide from both electrodes. The direction of current

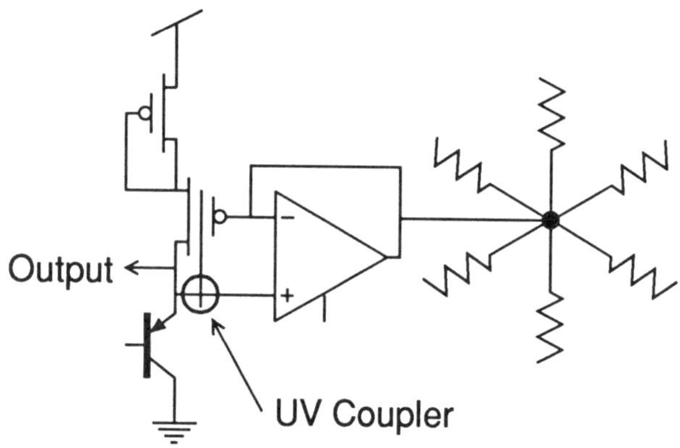

Figure 3 Schematic of a pixel that is similar to the one depicted in Figure 2, but that can be adapted with ultraviolet light to correct for output variations among pixels.

flow is determined primarily by the direction of the electric field in the oxide. In other words, the application of ultraviolet illumination to a capacitor with silicon-dioxide dielectric has the effect of shunting the capacitor with a very small leakage conductance. With no illumination, the leakage conductance is effectively zero. The leakage conductance present during ultraviolet illumination thus provides a mechanism for adapting the charge on a floating gate.

We can make use of ultraviolet adaptation in our high-gain retina using the circuit shown in Figure 3. This circuit is identical to that of Figure 2. except that a floating gate has been interposed between the resistive network and the pullup transistor for the output node. The network is capacitively coupled to the floating node. The current into the output node is thus controlled by the voltage on the network, with an offset determined by the charge stored on the floating node. There is a region where the floating node overlaps the emitter of the phototransistor, shown inside the dark circle in Figure 3. The entire chip is covered by second-level metal, except for openings over the phototransistors. The only way in which ultraviolet light can affect the floating gate is by interchanging electrons with the output node. If the output node is high, the floating gate will be charged high, thereby decreasing the current into the output node. If the output node is low, the floating gate will be charged low, thereby increasing the current into the output node. The feedback occasioned by ultraviolet illumination is thus negative, driving all output nodes toward the same potential.

Adaptation of the Retina

A scanned representation of the response of the adaptive retina to uniform illumination is shown in Figure 4. The dark bars at the left of each pixel are due to a capacitive transient as each pixel output is addressed. The hexagonal organization of the pixel array is clearly visible. In this and subsequent adaptation examples, the resistive network has been adjusted to have very low resistance relative to the transconductance of the amplifiers in each pixel. Under these conditions, the network computes a global average across the chip. Notice the wide variance in output voltage; many outputs are saturated positive, and others are saturated negative. We can adapt these nonuniformities by exposing the chip to uniform ultraviolet illumination. Figure 5 shows the response of the adapted retina after five minutes of such exposure. Notice that all pixel outputs are now uniform. The adaptation has completely eliminated offsets due to inhomogenity in device characteristics.

The response of the retina when two opaque bars are interposed between the chip and the light source is shown in Figure 6. No center-surround response is observed for this adjustment of the chip, because the network averages over the spatial scale of the entire chip. We can now perform an experiment that can also be performed on the human visual system. We can adapt the retina in the presence of an image fixed in the visual field. The result of adapting the silicon retina to the stimulus of Figure 6 is shown in Figure 7. The chip has generated an *afterimage.* The contrast of the afterimage is reversed from that of the image to which the retina was adapted. This kind of afterimage is produced in our own retinas if we stare at a fixed image for a long time. Our retinas are constantly adapting. The only reason that we are not constantly confused by such afterimages is that our eyes are constantly in motion. The illumination at any one pixel, averaged over many visual scenes, is the same as that for any other pixel. We can approximate the mode of operation of human retina by fitting the chip with a quartz lens and operating it at high altitudes, where enough ultraviolet light is present in the solar spectrum to adapt the circuits continually.

Once the retina has been adapted, we can increase the resistance of the resistive network to reduce the distance over which the lateral averaging takes place [Mahowald 89]. With an averaging distance of a few pixels, the retina shows a strong center-surround response. The output of the retina under these conditions is shown in Figure 8.

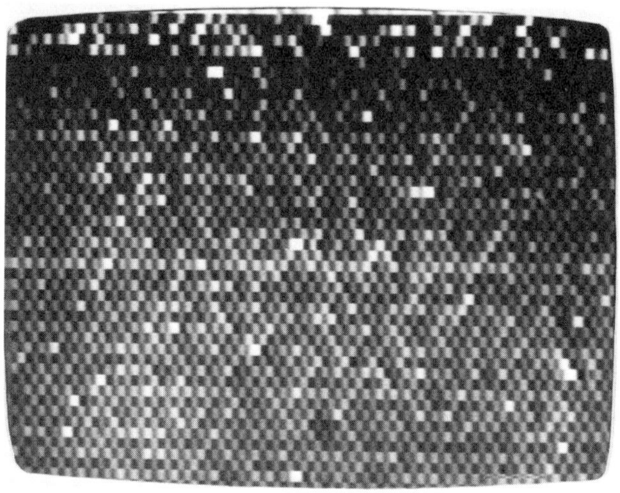

Figure 4 Scanned representation of the output of the adaptive retina, before adaptation

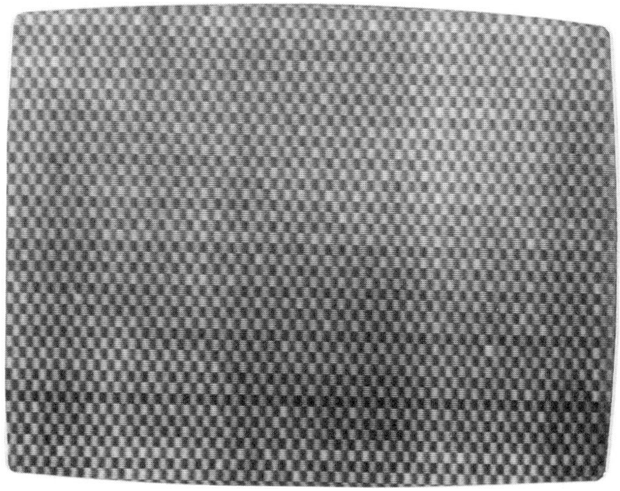

Figure 5 Scanned representation of the output of the adaptive retina, after adaptation

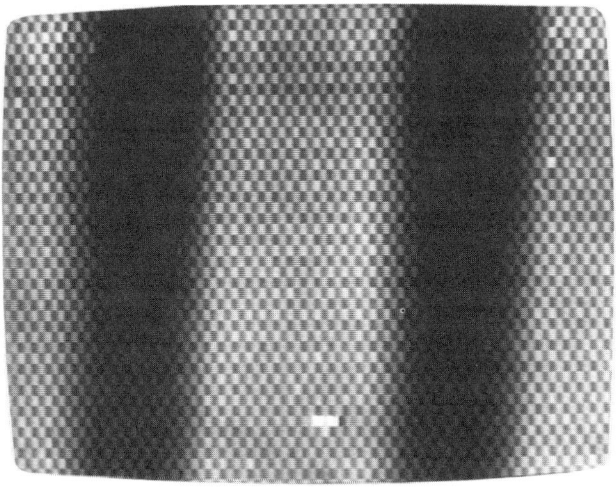

Figure 6 Scanned representation of the output of the adaptive retina with two opaque bars interposed between the ultraviolet light source and the chip

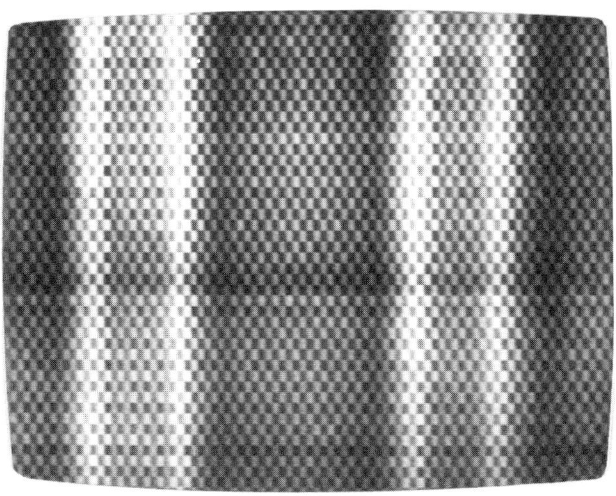

Figure 7 Scanned representation of the output of the adaptive retina after adaptation with the two dark bars of Figure 6. The negative afterimage is clearly visible.

Figure 8 Scanned representation of the output of the adaptive retina after adaptation, with the space constant of the resistive network adjusted to show center-surround response.

Acknowledgements

This work was sponsored by the Office of Naval Research and the System Development Foundation. Computation was provided by a grant from Hewlett-Packard, and Chip Fabrication was supported by DARPA, through the MOSIS service.

References

Glasser, L. A. (1985). A UV write-enabled PROM. In *1985 Chapel Hill Conference on VLSI*, pp.61–65. Rockville, MD: Computer Science Press.

Mahowald, M. A. and Mead, C. A. (1988).
A silicon model of early visual processing. *Neural Networks*. **1**:91–97.

Mahowald, M. A. and Mead, C. A. Silicon Retina. In Mead, C. A. (1989). *Analog VLSI and Neural Systems*, pp.257–278, Reading, MA: Addison-Wesley.

Mead, C. A. (1989). *Analog VLSI and Neural Systems*.
Reading, MA: Addison-Wesley.

INDEX